'This book has found a special place in m[...]
and unexpected, as precise and exquisite [...]
I loved it.'

Helen Macdonald, author of *H is for Hawk*

'Everyone should own *A Honeybee Heart Has Five Openings*, which moved and delighted me more than a book about insects had any right to ... Jukes is a gloriously gifted writer and her book ought to become a key text of this bright moment in our history of nature writing.'

Alex Preston, *The Observer*

'Finely written and insightful.'

Melissa Harrison, *The Guardian*

'Beautifully written and timely.'

Rob Macfarlane, author of *Underland*

'*A Honeybee Heart* is a window onto the world of the hive, its intricate strangeness and beautiful patterns. It's a tenderhearted account of life at its most tenacious and most vulnerable. Jukes' writing will make you see the details around you differently: you'll see new shades of blue and pollens and poisons and small acts of care everywhere. This book is a special kind of treasure.'

Naomi Booth, author of *Sealed*

'Lyrically beautiful.'

Philip Hoare, author of *The Whale* and *Leviathan*

'So beautifully written, it's become part of my life. Definitely one of my books of the year.'

Fiona Talkington, BBC Radio 3

'This is classic modern nature-writing; a synthesis of scientific learning, observation and the author's response. If you care for the wellbeing of bees and the planet and for the state of the human heart, then this book, with its deft and beautiful prose, is for you ... And like all good nature writing, it also – quietly, clearly and insistently – requires us, too, to respond.'

*Countryfile Magazine*

'An intimate exploration of the heart and home, and a tantalising glimpse into an alien culture. A brave and delicate book, rich and fascinating ...'

Nick Hunt, author of *Where the Wild Winds Are*

'[A] subtly wrought personal journey into the art and science of beekeeping. Helen Jukes evokes both the practical minutiae of the work, and the findings of researchers who have illuminated bee ethology over the centuries.'

*Nature*

'A mesmeric, lovely, quietly powerful book. A gentle but compelling account of the redemption that comes from relationship and attention.'

Charles Foster, author of *Being a Beast*

'I love suddenly finding a book you can't wait to read. It doesn't happen often . . . *A Honeybee Heart* is enchanting, real and full of insight.'

Mary Colwell, author of *Curlew Moon*

'A profound, funny and sometimes deeply moving book that describes a year of inner city bee keeping, while dancing between the history of bees and us and what it means to be human in our modern world.'

Julia Blackburn, author of *Threads: The Delicate Life of John Craske*

'A very human story about the aliens gathering in her back garden – bees, fascinating but almost unknowable. Their wildness and her duty to them help open up a desk rat's uninspiring life to all the possibilities of love, care, connection and sheer wonder. It is a lovely, entirely personal journey into the very heart of the hive.'

Michael Pye, author of *The Edge of the World*

'I raced through this really terrific, down-to-earth read. The existential threat to our entire ecosystem posed by the problems facing bees can be hard to grasp, but Helen manages to make this a very personal, human story that, hopefully, might inspire others to action.'

Luke Turner, author of *Out of the Woods*

# A Honeybee Heart Has Five Openings

*A Year of Keeping Bees*

HELEN JUKES

## SCRIBNER

LONDON  NEW YORK  TORONTO  SYDNEY  NEW DELHI

This paperback edition published by Scribner, an imprint of Simon & Schuster UK Ltd, 2019
First published in Great Britain by Scribner, an imprint of Simon & Schuster UK Ltd, 2018
A CBS COMPANY

Copyright © Helen Jukes, 2018

SCRIBNER and design are registered trademarks of The Gale Group, Inc.,
used under licence by Simon & Schuster Inc.

The right of Helen Jukes to be identified as author of this work has been asserted
in accordance with the Copyright, Designs and Patents Act, 1988.

1 3 5 7 9 10 8 6 4 2

Simon & Schuster UK Ltd
1st Floor
222 Gray's Inn Road
London WC1X 8HB

Simon & Schuster Australia, Sydney
Simon & Schuster India, New Delhi

www.simonandschuster.co.uk
www.simonandschuster.com.au
www.simonandschuster.co.in

A CIP catalogue record for this book
is available from the British Library

Paperback ISBN: 978-1-4711-6774-4
eBook ISBN: 978-1-4711-6773-7
eAudio ISBN: 978-1-4711-6897-0

The author and publishers have made all reasonable efforts to contact copyright-holders
for permission, and apologise for any omissions or errors in the form of
credits given. Corrections may be made to future printings.

Typeset by M Rules
Printed and bound by CPI Group (UK) Ltd, Croydon, CR0 4YY

## A Note from the Author

All stories are shaped in part by their telling, and it should be noted that – in the process of putting this one down into book form – some names and details have been changed.

# Contents

# 1

## *Doorway*

# November

The day the idea arrives I am wanting badly to escape. Home from work and too wound up to stay inside I open the back door, step out. There's a scuffed-up doormat at my feet printed with three faded purple owls and *Welcome* in big letters underneath. It's the wrong way round, it welcomes leavers; it'd be upside down if you were arriving here. I stare down at it and blink. A nerve at the back of my eye buzzes as if the whirr of the computer screen has got inside my head. My shoulders are hunched and my neck is stiff. A thick wad of muscle has bunched itself at the top of my spine and now I knead it with my knuckles, hard.

I'm tired. And I'm still wearing my work shoes, which are not made for walking about in a frosty garden, at dusk.

But this evening I need to cover some ground – to get somewhere else, not here. In the back garden of an end-terrace on a busy road leading out of Oxford's city centre you can only get so far. I count the strides, and make fifteen. Past the shed with a vine like a trailing wig and the pond silted with fallen leaves. Along the wall adjoining our neighbours' garden, which crumbles slightly when you touch it. Near the end of the garden this wall gives out altogether and becomes high beech hedge. Here is a compost bin, and then a thicket of weeds.

I moved in recently, with my friend Becky. I'd been offered a job working for a charity in Oxford just as the last project I'd been working on in Sussex was drawing to a close. It was a permanent contract, and after a lot of moving around over the last few years, that felt like an opportunity; a chance to stay in one place, maybe even settle down a bit. When I called Becky and told her I was moving to the area, she suggested we get somewhere together. So then we found this place. A red-brick two-up two-down with clothes moths in the carpets and a narrow garden at the back that's grown overcrowded with weeds. That was a few months ago, and it hasn't been an outright success so far. The job's been tough, and I've been struggling with the workload. Wishing I had a thicker skin, and was better at managing things like office politics and fluorescent light bulbs and those desk chairs with the seats that spin and spin. Last week, a colleague told me that both my predecessors quit when they hit overload, and it was clear from her face that she was not expecting the story to be any different this time around.

At the far end of the garden is a wooden fence. It's hidden behind a loping conifer and dried-up gooseberry bushes, hidden again under a mess of brambles, so you wouldn't know it's there or quite where the garden ends – except for a gap to one side, between a holly bush and a bird table, where you can see it. I squeeze through, and touch the fence. Tiptoe up, but I can't see over it. And now for one moment, maybe two, sheltered by the holly, which also pricks my thighs, I forget where I am. Forget the house that doesn't feel like home yet, and the hectic work schedule. This is when the idea arrives. Here is where the bees would be, I think, and then catch myself thinking it. Step back with surprise. It used to be a habit, looking for gaps like this. It's been a while since I remembered it. But now I begin checking for prospect, wind exposure, the damp. I glance up, to where the trees won't shadow them. There's a warehouse roof some distance away, the sun sinking. A plop behind me, as a raindrop falls.

\*

I learned a bit about beekeeping a few years ago when I lived in London, where I met Luke, a friend of a friend, who has hives all over the city. His beekeeping began as a hobby: he was given a small plot at the Natural History Museum in exchange for a pot of honey each year – but then it grew. Soon he was being approached by other companies who wanted to keep bees, and they were offering to pay him. By the time I moved to London and asked for an introduction he had hives at magazine- and fashion-houses, pubs, hotels – he was keeping the bees and training the staff until they could do it for themselves.

Urban beekeeping was still unusual at that time, and I'd never seen inside a hive. It sounded fun, and different, and – feeling dizzied by the scope and sprawl of my new home city – I was keen to meet someone with half an eye on the lives of small and humming creatures.

The first time we met, Luke was wearing a cream three-piece suit, a pink shirt and a summer boater, and he was swinging a blue IKEA bag. He exuded charm – '*Helen!*' he beamed when he saw me. 'How *wonderful* to meet you!' We were outside Coram's Fields, a children's park in central London, where he kept two hives in a thin strip of undergrowth behind the cafe.

'So you've come to see the bees?' he said, and I nodded. Underneath his hat was a head of short grey hair. He looked a bit like a mole, I thought, as I spied metal contraptions and gauze masks inside the bag. 'Some people believe that bees can smell your fear,' he said, as he unlocked a gate in the iron railings and we followed a gravel path around. So as we pulled on our suits I concentrated on not being afraid, but when he lifted a hive lid and they began seething out I was terrified.

I hadn't even realised until that day that honeybees are different from bumblebees; that there are over twenty thousand species of bee in the world, and only a small fraction of them make honey. '*Apis mellifera*,' Luke announced, as though introducing an old friend. That's the western honeybee, and the one most extensively kept and bred.

These bees were not fuzzy and they were not soft. They were brittle and trembling and when Luke lifted the hive lid they didn't buzz, they *hummed* – like a machine but more unstable, more liable to volatility. Beneath the lid the space was packed with wooden frames hanging perpendicular to the roofline, each one filled to its edges with comb covered and crawling with bees.

'Look,' Luke said as he lifted a frame out, pointing first to where the queen had laid eggs inside the cells, then to where the workers had stored pollen for feeding young larvae and finally to where nectar was undergoing its conversion to honey. Honeybees are among the few species of bee to live together as a colony – even bumblebees, who are social in summer, reduce down to a single queen in winter. They work to produce as much honey as they can while flowers are blooming so as to sustain themselves through the cold season.

They were crowding from the frames and from the entrance. We had unsettled them, and now they wanted to unsettle us in return. I glanced over at Luke, who was working calmly and swiftly, with an ease I hadn't noticed before.

'They're swarming!' I yelped.

'They're not swarming,' he said. 'Swarming is what happens when a colony splits and leaves a hive; these lot are just defending this one.'

I was hooked. By the bees, and by the beekeeping too – those precise and careful movements that were not unlike tenderness; not unlike a kind of intimacy. Soon I was beekeeping whenever I could. Luke would send a text message with an address and a time of meeting, and I'd jump on my bike and race through the streets to go and join him. It felt like slipping through a hidden side-door, stepping slightly outside the flow of things and into a different version of the city. Nothing was as it first appeared when we went beekeeping. Walls had recesses, windows could be climbed through, roofs climbed onto. We followed underground tunnels and hidden passageways, entered green spaces I hadn't guessed were there. But all of this was peripheral to the actual task of *opening* a hive, when we had to settle down, become very attentive to the colony and ourselves. The beekeeping suits covered us from hooded head to boot-clad ankle, and looked more like they'd been designed for protecting against nuclear radiation than opening a beehive. Inside the suit I was both cocooned and strangely conspicuous – that space behind the cafe at Coram's Fields bordered a pavement, and passers-by used to stop and point through the park railings as we worked. We hardly noticed them. Once the lid was off, we were absorbed. Each movement of arm, leg, hand and head was freighted – a sudden grab or drop would disturb the bees, and then we'd have to watch awhile and wait as the disturbance moved through the colony as a wave or a change in frequency or a shudder.

I could do with finding a hidden door now, I think to myself, crunching back over the frosted grass of the Oxford garden with my arms folded and my hands tucked into my armpits. Perhaps I *will* get bees, I think, looking up. And by the time I've reached the back door of the house the idea is already taking shape in my mind, gathering and becoming solid, bedding itself in.

*Yes*, I think, eyeing the collection of abandoned plant pots by the doorway. We could do with a bit of pollination here. Something to inject a bit of life. My fingers are like ice blocks and I'm not sure if they're freezing my armpits or if my armpits are thawing them.


Next day I'm at work again, pinned between a laminate desk and a wall.

The office is small. There are five workstations jigsawed in, each one a slight variation on a type. Desk, computer, chair, worker – like not-quite-conjoined cells, and you can't see who's inside each one except by leaning, which is a dangerous game when you're seated on a swivel chair with wheels. Each desk has a hole in the top the size of a clenched fist, through which a bunch of cables runs down into the floor. The carpet is made from nylon threads, like a head of hair that's been squashed down or sat upon – I know this because once when I was here and there was no one else in the room I lay down and looked at it close up. The walls are white. As I walk in each morning I feel the muscles at the back of my neck clench, and they stay like that until I leave.

It's late afternoon and my attention has strayed. The plant I brought in to brighten my desk has died, and I am unsure how to dispose of it. Outside in the corridor people are shuttling past, shoulders pinched, their feet thudding dully over the squashed-down carpet hair. A girl from the marketing department hurries in and dumps a pile of papers on my desk. 'You asked me to print the posters but I can't print the posters,' she says, loud enough that everyone else in the tight-packed room can hear her. She can't print the posters because the printer is broken, and the person who normally fixes it is off with stress.

We look at the pile of papers. She shrugs at me. Then she turns on her heels and leaves.

I shift the pile to the edge of the desk and blink at my computer screen. I want out, I think, then quickly bury the thought. Because I can't just get *out*. I've moved houses, changed cities, to take this job – I can't just up and *leave*.

The skin around my eyes is tight. Maybe the screen is too bright or my focus is too narrow or maybe the muscles are tired of bracing themselves against everything that has been pressing in. I rub my eyes, refocus. This is when the idea comes back again.

'I might get a beehive,' I say out loud, to no one in particular.

Joanna who sits at the desk opposite bellows a laugh and gestures over to my sorry pot plant. 'Madness,' she says bluntly, and returns to her screen.

No one else says anything. I don't say anything.

And that, it seems, is the end of it.

*

But I forget that words are important. Once you've gone and said a thing out loud, people start holding you to it. Once you begin describing something in your head, you are already setting it in motion.

Later that week my friend Ellie comes over with a bunch of grapes and a box of peppermint teabags. Ellie is almost a whole head taller than me, with long dark hair that always looks windswept. If she were an animal she might be a hare, because of her long legs and her eyes that are wide and green and rimmed with black. She has a love of language so precise and clear that sometimes when I listen to her I think it doesn't matter what happens; everything is okay, as long as you can find a way to describe it.

'How's things?' she says, as we stand beside the kitchen counter and wait for the kettle to boil. The grapes are small and green and their flesh is so tart inside their tightened skins that we wince with each bite into them.

'Okay,' I say. 'Busy. I've been thinking of getting a beehive.' Throwing a grape up and catching it between my teeth, as though getting a hive is a thing that one might quite casually fall into.

In fact, I haven't stopped thinking about it. Over the weekend I hunted through cupboards for my old beekeeping suit, which I found in a trunk under a pile of curtains and dried-out mothballs. I've hung it from the banisters to air. And now whenever I walk upstairs the top of my head brushes its sleeves.

'I didn't know you kept bees,' Ellie says, leaning over the sink to get a better look at the garden. 'Could you have them here?'

'Oh yes,' I tell her. 'We're bordered on two sides by vacant land. It's just wild space over the fence there, and allotments behind. And that beech hedge is screening the neighbours' garden. It's perfect, really.'

She leans further and looks. But it's been raining all day, and everything is sunk into the same dull grey as the traffic and the street beyond. It's difficult to imagine there being enough here to sustain even a single bee, let alone a whole colony of them.

So instead we begin a kind of game, imagining the kind of person who becomes a beekeeper; sketching a character in words:

'It's a man.'

'Yeah, it's definitely a man.'

'Likes his own company.'

'On the edge of retirement.'

'Retired already.'

'Or with a lot of free time.'

'And a sweet tooth.'

'And a thick skin.'

'Or a thick suit.'

'Keeps his calm in the midst of a swarm.'

'Uh-oh,' I say, and make a face. I am not always good at keeping calm.

I'm still in my work clothes, thin tights and pinching shoes and a dress that itches at the knees. We could go on. But the more we do, the more remote I feel from actually becoming a beekeeper myself. The idea that had begun taking root in my head begins to twist, contorting itself until I wonder if getting a hive really is a kind of madness. Aren't I supposed to be heading out of the front door, getting to know the area, rather than hiding away by the far fence?

'I saw a film last year about commercial beekeeping in America,' Ellie's saying. 'It's not just about honey any more; some of them make half their income supplying pollination services in places that have been over-farmed. They drive around in massive trucks, hundreds of hives on the back. Migratory beekeeping, it's called. The film followed one guy who spent most of the year on the road.'

'I remember a news story last year about one of those trucks,' I say, pouring tea and passing her a mug. 'A big crash, I think it was Washington State. Four hundred hives fell out and scattered across the road. It was just before dawn. The sun came up, and swarms of bees came with it.'

'What happened?'

14

'To the bees? Firemen came, and hosed them down. Killed a lot of them. I think beekeepers from a nearby town arrived and caught some.'

'There was a bit in that film about China,' Ellie says, cupping her hands around the mug. 'Places where there are no bees left. The owners of apple orchards were employing people to do the pollinating. They showed a clip of it. Farm labourers up trees, clambering around with tiny paintbrushes. Bodies half-hidden by the blossom.'

On her way to the loo she passes the suit hanging over the stairs. 'Wow,' she says. 'You're really thinking about it.'

I look up at the suit arms dangling and their elasticated cuffs and realise suddenly how different it would be to keep a hive of my own. When I was beekeeping before it was always under Luke's guidance. All observations drawn, all decisions made, could be checked and passed by him. We were moving between places, covering distances. I never quite took responsibility, I was never really accountable to a single hive.

It would be another story to have a colony of bees here, outside my own back door. And me dressed up as the keeper of them. In truth, in the rest of my life I'm not so well versed in *keeping* things. Like a lot of my generation, I've moved around a lot. I haven't lived anywhere longer than eighteen months in the whole of my adult life. I'm thirty. I don't find it difficult to make new friends, but bar a few misjudged kerfuffles, I haven't had a boyfriend for years. Restlessness doesn't sit easily with intimacy or is a neat way to avoid it, and there is something a bit terrifying now about the thought of inviting another creature in.

There does seem to be something about beekeeping that *gets in*. I've met beekeepers who talk about *getting the bug*, that it *gets under your skin* – as though bees had become an obsession, and keeping them a compulsion. Luke doesn't talk like that, but he did tell me once that soon after he started beekeeping, he began noticing colour differently. Bee eyesight spans a different part of the colour spectrum from ours; they see more blue than we do, and many insect-pollinated flowers have evolved to bloom in blues and purples so as to make themselves more visible to bees. Luke found that he too had begun noticing blues around him. Not just the lavender bush outside his door but also the napkin beside his plate, that man's watch-strap, this bottle top skittering over the pavement. He'd been drawn towards colour, and also drawn into the uncertainty of changing weather patterns. 'If it rains, my plans change. The weather decides if I can go beekeeping, not me.'

'It's been a strange year,' he said to me one spring, after

it had seemed to rain for a month, and then was hot, and then snowed. I also began noticing things I wouldn't have otherwise. Stepping out of the suit, unfettered by any real beekeeping responsibilities, and returning to the city streets, I'd see details I'd overlooked before; I became aware of what creature life existed behind and between London's walls. Sometimes my eyes felt altered after opening a hive. I wondered what it was about the bees that made the looking different.

Later that evening, when Ellie has gone home, I'm in the front room and clambering over the furniture.

'*Helen*,' Becky says, arriving back from work, and ready for food and bed. 'There are no clean pans left. Are you going to finish that washing up?'

'Yes,' I say, but I don't make any moves towards the kitchen. I'm balancing with my toes on the edge of an armchair as I reach up for a bookshelf. 'Have you seen my bee books?'

I tried to get a good look at the bees in Luke's hives, but a bee is not an easy thing to observe. The thing is that they are so *strange*. In just the same way that you can think of light as either a wave or a particle, a honeybee can be thought of as either an individual or as a single cell within the larger superorganism of the colony. That disrupts our notion of the world as a place of fixed forms, where every creature has a name and a set of identifiable parts. You might look at a dog or a cat – the arrangement of eyes, nose, mouth and ears – and you'd be able to recognise and relate to a face. But with a bee you'd need a microscope, and even then the body is so strange that you might have to reach for diagrams to make sense of her. And what about a colony? A colony is nebulous and shifting; when it takes flight as a swarm it can seem to belong more to the air than anything in the physical world. You can't draw a ring around it; can't make a body out of it. So how to get a look? How to go about understanding it?

With so many tight deadlines and tick-boxes at work, perhaps I've become a bit black-and-white recently, needing to define things as either one or the other, never both or between. But maybe with the bees you have to adjust your senses – to slip between habitual understandings of eyes and nose, mouth and ears; of face and body, form and formlessness – to be able to *see* them.

The books are up on the highest shelf, and I have to stretch to bring them down. One of the things that happened when I was beekeeping before was that I began reading about bees, and now I feel it coming back like a thirst. That feeling of wanting to *know* the hive – to ask questions and find out and understand. I lay the handful of books out like stepping-stones across the floor. They reach as far as the doorway, and Becky's feet. I look up and remember the pans waiting.

A few days later I'm out in the garden with an old ice-cream tub full of orange rind for the compost. If you dump kitchen scraps over an open heap like this in summer then a cloud of flies lifts up, but in the winter nothing moves. I empty the tub and the rind tumbles out. The heap has an orange-crested top.

'*Psst. Hello!*' There's a sound of dry twigs cracking, and I jump because I'd thought I was alone out here. I turn round to find my neighbour crouching down on her side of the hedge. The hedge is high, and we're not normally overlooked – but now the leaves have fallen and the whole thing is bare enough to see through.

Her name is Hannah. She's young – in her early twenties – with extraordinarily straight brown hair that never seems to get longer or shorter. She cycles to work with her fiancé every weekday morning at 7.55 a.m. and they arrive back at 5.35 p.m. They have mountain bikes, and matching cycle suits.

'All right?' she says. And then, 'Darren told me about the bees.' Darren is her fiancé. Apparently Becky has already mentioned to him that we're thinking of getting a

19

hive, and Darren has already approved. 'Where will they go?' she asks, her voice bright and light, and I point to the fence, that space by the holly, obscured by a lumbering bramble bush. 'And when will you get them?'

I begin willing the hedge to cloak itself with leaves again, give me back my rim of cover. The hive is supposed to bring a bit of respite; I'm not ready for questioning yet.

'I don't know,' I say. 'Spring, maybe.'

And then she shuffles closer so that the beech twigs must be prodding her cheeks. 'They're in trouble, aren't they? The bees. I was reading an article the other day about declining honeybee populations. No single cause, the article said. A mixture of *in-ter-rel-at-ed* factors.'

She says *interrelated* carefully, sounding out one syllable at a time, as if she is still fitting the word together inside her head. And it's true. Pesticides outside the hive, parasites in; shifting patterns of land-use, changing climates and the weather – the list of factors affecting the hive over the last fifty or so years goes on and on.

'Gave me the shivers,' Hannah says. 'What do you think?'

'About the decline? I think it's complicated. Imagine all those stressors building up inside your house; something's bound to snap after a while.'

She's still crouching down so I crouch too, and then our eyes are level with each other among the beech twigs.

'Maybe they're like canaries,' she says, twisting a twig until it snaps. 'You know, the ones that used to get taken down the mines. Their lungs are so small they'd drop down dead if there were poisonous fumes around, and then the miners would know to get out quick. Perhaps the bees are trying to tell us something.'

'Yeah,' I say, remembering how I'd read that in myths and folktales from cultures across the world there are stories of bees as messengers, able to pass between realms. In ancient Greece the sound of bees buzzing through the cracks of rocks was said to be souls emerging from the underworld; the Mayans believed that bees were imbued with mystical power; and in British folklore they're known as small messengers of God. I think of the miners in the deep dark tunnels who turned round to find their bright yellow birds had snuffed it. 'Perhaps,' I add. 'Except that we can't go up for air.'

'No,' Hannah says. 'Right.'

She straightens up and dusts herself down. 'I'd better go. Good luck with it.' I think she's about to leave, but then she pauses a moment. 'Oh, and watch out for the citrus.'

'Sorry?'

'The citrus.' She nods at the ice-cream tub which I am still holding in my hands. 'Too acidic. It's not good for compost.' I turn and stare at the heap, which looks like it hasn't done any decomposing for months. 'You need a lid,' she says. 'Something to keep in the heat.' And I hear a soft tutting as she walks back up to her house.

Next week the weather gets even colder. The city canal freezes and part of the river does too, and in the mornings I walk to work under a sky flushing the same cold pink as my skin. Front gates, car wing-mirrors, the spindly spine of a rose bush – everything is sharp and hoar-spiked.

It doesn't stay that way for long. Soon the clouds roll over and the temperature rises, and we get a lot of rain. Spots of mould appear on my bedroom walls and books. Furred, and forming in perfect circles. I trace their perimeters with the tip of a finger, rub them off with an old tea towel. The mould disintegrates. It is like dust. It leaves a blue-grey smudge on the fabric.

I've heard that if there are mould spores in your house they start growing inside you too. So, lying in bed at night, I begin imagining the organs and routeways in me growing furred. Lung, trachea, larynx, throat. Lungs are interesting. With that passage of air moving in and out they're our only internal organs to be in constant contact with the outside world.

And, lying in bed at night, I also get to wondering whether I'm up to this business of getting a hive. What if they die or swarm away, or what if I want to get away from them? What if they keep me here, stuck in a place where I feel so unsettled? What if I can't keep them?

I stopped beekeeping before because I was set to move again, and out of London this time. 'But you were just getting the hang of it!' Luke said, which surprised me because I hadn't thought I was getting the hang of it at all. I still hadn't *understood* the bees, and if there was an internal logic of the hive I hadn't learned it. That winter we'd lost the colony at Coram's Fields, and I was still reeling from the sight of those greyed and lifeless bodies all huddled together on the comb. I obviously hadn't done anything to *help* the bees, and all this time spent beekeeping was becoming difficult to explain to other people. London is so full of drive and push and moving up-up-up that you can feel like you're impeding the current if you're not trying to get anywhere in particular.

Outside, in Oxford, it keeps raining. I have a big report to write and for a couple of days I work from home, so I'm in the house during daylight hours more than usual. Whenever I look out of the back window it is that space by the far fence that I see first. There's something waiting, or watching, about it. It makes me itchy. I look away, back into the room where my laptop is sitting, seventy-two work emails waiting. Seventy-three.

A muscle at the back of my neck twinges. Some hard and determined knot has wedged itself deep into the base of my neck, and now it's getting aggravated. A shoot of pain runs from my head down into my wrists, humming there a moment, and then it's gone.

I put up a hand and rub the tops of my shoulders, trying to find and loosen the knot, but I can't – it's driven too far in. As I am rubbing, the rain is tapping like fingers, or tiny, toughened bodies, at the glass.

And then the rain stops and my friend Jack arrives, with pruning shears. Jack lives down the road. He's over six feet tall and looks a bit like an enormous seal, with great dark eyelashes and a body made to be in water. A section of the Thames runs through the city half a mile from his house, and he swims there every unfrozen day of the year, in his pants. Wherever he goes, he always takes a towel. If he comes across water but has forgotten his towel, he swims anyway.

'Pruning shears?' I say, bewildered.

'Thought you could do with a bit of help clearing the garden,' he says, shifting his weight from one foot to the other. 'Get it ready for the bees.' Jack is another person I've told about getting a hive, and Jack doesn't wait around. I stare at him a moment, unsure whether I want to clear the fence or just leave it as it is, inhospitable to human or hive and tangled up with weeds.

'But everything's still wet from the rain,' I say, eyeing the shears.

'They're rusty already,' he shrugs, pulling them open and closing them. They make a harsh grating sound with each movement out and in, as if to demonstrate their unsuitability for the task.

I reach for my boots. I find us some gloves, too.

'What a mess,' he says when we're down by the fence. And then he begins cutting the brambles and trimming the holly. The shears are very blunt. It is more like hacking than trimming, but it works, and the clearing by the fence expands, and after a while I even begin enjoying myself, hauling armfuls of dead twigs back and forming a small mound of them on the grass.

Two round sweat marks appear on either side of Jack's shirt. 'There's a Lithuanian word for friend,' he's saying over his shoulder. '*Bičiulis*. It comes from the Lithuanian word *bitė*, the word for bee. Literally, it means *a person with whom one shares the keeping of the bees*.' Jack has an enormous capacity for random facts. Sometimes he has trouble getting to sleep at night, and he uses these periods of extended wakefulness to teach himself new things. Last month it was home brewing. Now it seems to be Lithuanian.

'Huh,' I say, tearing a band of rotten ivy from the fence and wondering how to let him know that this idea of getting a hive isn't really about friendship; it is just about me and the bees, and I'm not sure I'll want anyone around to watch.

'Yeah,' he's saying. 'In ancient Lithuania bees weren't bought and sold, so relationships between beekeepers were based on giving and borrowing. They linked people across whole districts. I guess beekeepers must have been thought of as honest, trustworthy people. Dear friends.'

I've heard of those relationships based around giving and borrowing. All across northern Europe there were traditions linking the buying and selling of bees with bad luck, whereas swarms that had been freely given or arrived of their own accord were believed more likely to do well.

Jack turns round. He looks at me and frowns. 'What *are* you doing?'

I'm standing by the fence, pretending I'm a hive, or a bee, testing this place for suitability. The brambles must have been growing here for years. The fence is scratched and softer and differently coloured in the spots we've cleared, and the earth is bare – just a few yellowed stems poking through, suddenly exposed to the wintry light and air. It's sludgy and uneven, too. A beehive should be placed on level ground, and this ground isn't.

When a colony of bees enters a cavity in the wild they

move up, around the edges, searching for the ceiling, a secure surface to build downwards from. Attaching themselves one to the next with tiny hooks on their legs, a group of workers will then form a loop, like a living chain. Inside this chain, more workers will gather and begin building wax comb vertically, along gravity's pull – a two-sided grid, with each hexagonal cell sharing its six sides with six adjacent cells. If the cavity or hive is crooked the comb will form at odd angles to it, and end up colliding with the walls.

From the garden we can hear the traffic rumbling along the road outside, splashing in the places where the storm drains have flooded and the rainwater has collected in great pools over the tarmac. A man shouts. A kid bawls. We stare down at the soil.

'It's not level,' I say to Jack.

'We can make it level,' he says.

'How can we?'

'I'm not sure. I need to think about it.'

We stand and look. The mud is seeping through a hole in my boot, dampening my feet. I try to imagine the inside of the hive, those first moments when the bees are in and searching around for a secure ceiling. What it would feel like if 'up' was where you fixed your feet, and depth what you built into.

I tell Jack about the honeycomb and the living plumb

line and he nods and smiles as though a hunch of his has been confirmed. 'Friendship,' he says, as though that settles it. 'They link up. They have to. It's the only way to get things done.'

# December

The first week of December, and any self-respecting bee must have retreated far inside her hive by now. Honeybees form a tight cluster in the cold just the same way that penguins do, circulating heat throughout the colony by keeping a steady flow of movement going, pushing those at the centre out and pulling outliers in.

The thermostat at work must be stuck on high, because everyone's walking around in T-shirts. With all the windows closed to keep out the cold the place is stuffy, and stuck at a drab mono-temperature. People keep getting sick. Yesterday someone blamed the closed atmosphere inside the office. 'It's a breeding ground for germs,' they said, and I thought at least there's something that's able to thrive in here. It can feel strangely airless – sometimes I get short of breath. I have to remind myself that the air is not missing; it just isn't moving very much.

At home we have the opposite problem. The central heating is on the blink, and a lot of heat has been escaping through the bare floorboards and the draughty windows and doors. We've been keeping the fire going most evenings, and actually I think I prefer it. With the embers and flames glowing the source is easier to locate, and the warmth always seems warmer that way.

Sprawled out in front of the hearth, I've been getting into my bee books. Tonight I'm reading up on what makes a good place to put a beehive. There's a tray beside me with a tub of gently melting butter and a half-empty honey jar. The tray is scattered with crumbs, my books and laptop too.

The Roman poet Virgil wrote about bees in his poem the *Georgics*. On the correct placement of a hive, he wrote: *First seek a settled home for your bees, whither the winds may find no access . . . nor straying heifer brush off the dew from the mead and bruise the springing blade.* So our little garden with a few frogs and birds and plenty of wind-shelter from the trees is doing okay so far. *But let clear springs be near, and moss-green pools, and a tiny brook stealing through the grass . . . In the midst of the water, whether it stand idle or flow onward, cast willows athwart and huge stones, that they may have many bridges whereon to halt and spread their wings to the summer sun.*

I like the idea of filling the garden with tiny bee bridges, and make a mental note to float some sticks in our pond come spring. But Virgil doesn't have much to say about terraced houses and rush-hour traffic so I put him down for now, and see what the government have to say.

The DEFRA website states that hives *should be sited so that only the beekeeper is ever likely to be stung.* Which seems a tricky thing to factor for in a city, and anyway I'm as keen to avoid stings as the next person. Yet hives are increasingly common in cities. Urban beekeeping is a big thing now – part of a resurgence over the last decade in beekeeping as a hobby. People with no previous experience have been starting one hive, maybe two, and taking a little honey each year.

The trend must be due in part to a growing unease provoked by the news of declining honeybee populations across the globe. Writer and academic Rebecca Giggs notes that one possible psychological response to the apprehension of a threat is to begin producing idealised versions of the thing we perceive as being at risk. Unable or unwilling to process the loss, we increase the intensity, brighten the colours, formulate for ourselves an ideal experience of the thing we believe might be slipping away. It's easier to ward off an anxiety over declining bee populations, or changing climates, if there are bees buzzing around. 'Look!' we can say. 'They're right here – everything's *fine!*'

But getting a hive won't save the bees, who don't and never have needed our keeping. If we want to do something to help them we'd do better to turn our attention to flowering habitats beyond the hive, on which they do depend. Wetland, woodland, wildflower pasture – all have been diminished over the last century in a hot wave of intensive agriculture and urban sprawl.

I scoop the sticky crumbs from the books and floorboards, return them to my plate. So what is attracting us to beekeeping, if it's not directly about saving the bees – and how do we become better custodians of their future? Perhaps it's unfair to see the resurgence only as another form of avoidance of what's *really* happening outside, to our climate and landscapes. Isn't it possible that there is some experience we're seeking, which we believe we might access, by bringing ourselves into an encounter with a hive?

The British Beekeepers Association describes beekeeping as a *therapeutic pastime*. It sounds a bit like a tagline from a tourist brochure – *Get closer to nature!* – and it makes me cringe. Our species has destroyed their habitats, moved inside the hive to manipulate their most intimate processes (you can now order an artificially inseminated queen, her wings clipped – and she'll arrive by airmail); it'd seem a little ridiculous to now turn round, seek nature out, and expect it to make me feel good.

But beekeeping is about more than gaining proximity to a hive; it's about entering into a *relationship* with a colony. So now I try a little exercise in my head. I try stretching my understanding of the word *therapy*, pulling its meaning at the edges to see what it can hold. And I wonder about a definition that places a different emphasis on *feeling better* – not feeling *better*, but *feeling* better: an approach to beekeeping focused on paying attention, becoming more attuned to the world around us, perhaps even adjusting how we sense and see. This might not always feel good, since it is not nice to see a colony perish. But if therapy is about gaining a more rounded perception of ourselves in relation to the world around us – how we affect our environments, and are affected by them; if it is concerned not simply with the business of tending tired egos, but the slower and more effortful labour of creating more sentient, compassionate and capable human beings – well, then perhaps lifting a hive lid and taking a look inside wouldn't be such a bad way to start.

I pick the honey jar up from the tray and turn it around, reading the label. Even this talks about *health-giving properties* – it seems you can't get anywhere near bees without some mention of healing. And who am I to question it? All those layers of history and meaning don't fall away because a marketing campaign got hold of them. Something *happens* when we come into contact with a hive; it's been happening since as far back as beekeeping began. Bees do something to us – they capture our imagination and beliefs, and our feeling.

They captured my imagination, for sure. When I was beekeeping with Luke he'd speak as we worked – naming things, telling stories – so that slowly my sense of the bees' world expanded. I began to enjoy the tang of fear I felt as we opened the hives, which gave way each time to a kind of intimacy or a quiet. In the city I was surrounded by people every day – crowds of them, wherever I went – but somehow it was by the hives that I felt most human. The colony was strange – wild and restless, and following a logic I could sense but never understand completely. Fascinated by the hive but often clueless as to what was actually going on inside it, *getting closer* in this context had added its own particular kind of complexity.

There's a clatter in the kitchen, and a shout. Becky's uncorked a bottle of homemade elderberry cordial, and it's exploded everywhere. She's in the doorway now, her face spattered with red juice. 'You thirsty?' she says, and grins.

*

'Have you opened this one?' It's Christmas Day. My brother reaches under the tree for a small handmade envelope and holds it out towards me.

'Who's it from?' my mum asks, leaning over the sofa and my dad.

'Becky,' I say, remembering how she'd slipped it into my hands as I left the house two days earlier, juggling gifts and wrapping paper. 'I think it's a Christmas card.'

But it is not a Christmas card, or not really. Inside the envelope is a pencilled drawing of a honeybee. On the back are the names of friends in Sussex and London and Oxford and beyond, who have all somehow got in touch with each other and put money in to buy me a colony of honeybees. Next to their names is the address of a farm near Banbury. *This is where the bees are waiting*, the note says. *You can collect them in the spring.*

'What is it?' my dad says, and they're all staring, a little bemused, because I must have turned white or started gaping.

'Bees,' I say, holding up the picture. 'I don't know how they . . .' How they all got in touch with each other, I want to say. What made them do it, why.

'It's all the excitement,' my mum says. 'It's been a difficult few months. She's very tired.'

And I'm touched, and thrilled, and panicked. I'll really have to get a hive now. A colony both bought and freely given – I'm not sure if that makes it lucky or not, according to the old folklore.

The journey back to Oxford on Boxing Day is slow. I take a late-night rail-replacement bus, and the bus gets stuck in traffic. Rain plastering the windows, two seats not quite big enough to lie along, even if I curl. Outside in the street, Christmas partygoers are returning home, and I spot three Santas with a crate of beer between them. They're shouting and jumping, trying to catch the attention of a Christmas fairy. She has a flashing tiara and furry silver wings and is doing a good job of pretending she hasn't heard them.

I look back down at my phone. I'm online, reading about a study by researchers at Sussex University, who used hives to monitor ecological health in areas that had been intensively farmed. Following planting schemes aimed at increasing biodiversity, they installed hives and recorded the types of pollen and nectar collected by the bees, thereby learning which flowering species could be effective in improving the diversity of the region as a whole. This is hive as barometer, as weather vane. We gain a sense, from looking in on this small place, of a whole landscape.

The phone beeps. It's a message from Jack to say he's

dropped two concrete paving slabs at my house. Found them in a skip; from an old doorstep. Thought they might make a good base for a beehive. Carried them down the hill in his rucksack. A weight. I look out, and the rain streaks. So now I have something to put it on, the hive. And maybe I'm going to need a few friends around after all. I imagine the sludging earth, the slabs sitting down over it. Try picturing the bees like a lightness above.

I can't imagine them yet.

2

*Hive*

# January

A new year begins. Before returning to work after the Christmas break I make a trip to London. I'm at my friend Dulcie's house in Forest Hill, sitting with her daughter. Her name is Corinne. She's four years old, and likes to draw pictures.

'A house,' she says, throwing her felt-tips down. 'Look!' She holds the paper up in front of her face, forgetting that just because she can see the picture, it doesn't mean that I can too.

'Come here,' I say. 'Let me see.'

A square shape with a rectangle inside and a triangle on top. House, door, roof; our first symbol for *home*. In Corinne's picture there is no ground below, or sky above. It is a house floating in space.

'It's good,' I tell her. 'Does it have any windows?'

She thinks for a moment. 'Behind,' she says. 'At the back.' And presses her chest against my knees.

Our template image for the modern beehive follows exactly this pattern. A square shape, with a triangle roof. A small entrance at the bottom, and a ledge like a doorstep.

The modern hive first came into use in the nineteenth century, and revolutionised beekeeping. The basic structure is composed of a set of wooden boxes stacked one on top of the other, each packed with a series of rectangular frames hanging vertically and in parallel, on which the bees build their beeswax comb. With a lid on top and those removable frames inside, this new kind of hive made it possible for beekeepers to carry out regular inspections of the colony for signs of disease, and to extract honey without harming the bees. Until then most beekeepers had used traditional wicker, clay or log hives, and common practice had been to destroy colonies at harvest. Often they were smoked with noxious fumes; sometimes they were drowned. Either way the bees were often killed, so as to allow the keeper to reach into where the honey was waiting.

With the modern hive, beekeepers could begin influencing the internal functioning of their colonies. The interior could be monitored, even tweaked – component parts could be added, removed, reworked to increase honey yields. Beekeepers learned to split colonies and breed them artificially, developing a new confidence in their ability to sustain and manage stocks. Dependency on the availability of wild colonies seemed a thing of the past.

On the way back from Dulcie's house I sit on a crowded commuter train and think about the endless adjustment and modification permitted by the modern hive; the will towards ever-increasing productivity. Don't the pre-occupations begin to sound a bit familiar after a while? Don't they seem to infiltrate everything? As though that impulse for more, and faster, and modifiable has recon-figured not only our hives, not only our places of work, but also our homes, our inner lives, our minds. *What's happening out there is also in here*, wrote Rebecca Giggs. We are the by-products, the weather vanes, of the times we live in.

I rub my neck and stare dimly at the reflection in the window of the man next to me, who seems to be falling asleep. It must be the time of year for booking holidays, because in the advertising space above our heads a travel agency has bought up almost the whole carriage. There are couples lounging on deserted beaches; families heading out on safari. They're all touting words like *getaway* and *escape*. Above the sliding doors an ad for a new smartphone promises a *life companion*. For a *richer, fuller, simpler life*, it says. Which seems like a contradiction in terms.

There are people standing in the aisles. Someone whispers a joke to the person next to them; someone else laughs. Pages of newspaper are being turned. A woman with a pregnant belly and a badge on her chest saying *Baby on Board* shouts 'Isn't anyone going to let me sit down?', and a man moves quickly for her.

We had a lesson at primary school once on tessellation. The teacher called it a maths lesson, but it seemed more like making pictures to me. Tessellation is the process of arranging shapes so that there are no gaps or over-lapping edges. We were given sheets to colour with the patterns all laid out ready – I had one with hexagons, like honeycomb.

I sometimes think that life must be a bit like tessellation for some people. You take one shape and fit it to the next and they sit comfortably together – you don't mind a bit of repetition because it's what makes the pattern form. Life is not like tessellation for me. Sometimes the shapes don't fit, or I don't fit into them, or I'm looking at the patterns but they don't feel real or right to me.

Back in Oxford after the trip to London, I get to

thinking about hive shapes. Now I know there's a colony of bees actually destined for my garden, I had better find something to put them in. But do I have to get a modern, conventional hive? Or might there be different shapes to work within – other ways of doing things?

I fire up my laptop and begin looking into what's available. Within a couple of hours I've gathered umpteen recommendations from umpteen people – websites, books and blogs – and they all tell me different things. Which is the best hive and how best to keep it. The entrance should be at the side, in the middle, and what about the *roof*? There should be a slanting roof or there shouldn't. A slant will leave a void inside, and then the bees will fill it. They'll build up into the cavity and the roof will stick, you won't be able to get in. But if the roof is flat then what about the *rain*? The rain won't run off, and imagine the *puddles*; imagine the damp sinking in from outside.

Before long, I get one of those overwhelming feelings that seem to come quite often from spending time on the internet, when you realise how much information is out there and yet how little you know. It's a bit like looking at the stars and realising how small and insignificant you are, but it is not so romantic and has absolutely nothing to do with awe.

So. Rather than make a decision I begin wandering

around on the screen, clicking links, scanning the range of beekeeping tools and equipment available. I order a *hive tool* and a *bee brush*, which feels easy and unintimidating enough. A hive tool is a flat, narrow piece of metal, about thirty centimetres long, with a chisel at one end – you use it to prise things apart inside the hive when they've been gummed together by the bees. A bee brush is for sweeping them away when you need to – I pick one made from *6-cm soft pure pig bristle*. I didn't know that pig bristles could even grow to six centimetres.

I also discover it is possible to buy sheets of factory-produced wax foundation comb to fit inside the wooden frames – like a base for the bees to build out from. Wax comb is used to house the eggs and young larvae known collectively as *brood*, as well as to store pollen for feeding the young and nectar for making honey. The foundation sheets are intended to give the bees a head start – with less comb to build, so the idea goes, they can put more energy into the important business of making honey. The sheets even come specially designed to manipulate the ratio of females to males within a colony. Left to build from scratch, bees manage the ratio themselves by adjusting the size of the comb cells. When the cells are smaller, the queen (the only fertile female in the colony) will know to lay a female egg; when they are larger, the queen will lay a male. By adjusting cell size artificially beekeepers can push the ratio further, upping the number of female workers and thereby increasing the productivity of the whole.

It's beginning to sound less like keeping a colony of bees, and more like keeping oneself in honey – and it's enough to make me uncomfortable.

'Want a cup of tea?' Becky pokes her head in through the door of our study-cum-dining-cum-everything-else room, and I look up in surprise.

'Yes – I mean no. Thanks.' Two hours have passed. There are thirteen or fourteen tabs open on the screen, and none of them are websites for hive stockists.

'What are you looking at?' she says, coming into the room.

I've been poring over diagrams of early hive forms. There are coil baskets, clay vessels, a *skep*. A bunch of reeds woven into a hat shape, plastered with dung, puckered at the entrance with handprints. Clay pots, upturned, their lips pushed flat and hard against the ground with a crack near the base for the bees to come and go from.

Wild honeybees nest inside naturally occurring cavities, and the practice of honey-gathering – finding these spaces, which might be a hole in a cliff face, or an inner recess inside a cave – was one of the earliest human activities. Beekeeping was different. This began when hands moved inside these spaces, feeling to the edges of them. Fingers smoothing an empty tree hollow, working at it, making it bigger, filling the gaps where a raiding wasp might slip inside.

Soon these spaces became objects in themselves. They were found objects at first: a hollow log, picked up and strung from a tree branch. Cavities became containers, something people could place in relation to themselves, and the bees came closer. There are records of organised systems of apiculture that date back to ancient Egypt. Paintings of beehives on the inside of temple walls, 4,500 years old.

They might have heralded the beginning of a new social order, but the first hives were crudely built. Twisted, smoothed and shaped from whatever materials happened to be lying around, and different across different regions. In her book *Bee*, Claire Preston describes how over the dry and arid areas of the Middle East, North Africa and Southern Europe the earliest hives were horizontal, the comb inside forming in rows along their length; whereas in the forests of Northern Europe where wild bees nested in hollow trees, the hives stood upright – as though each one took something of the landscape they were built into. It is all a long way from the hives for sale online today, made in one place, shipped to another, and replicated over and over.

Becky peers over my shoulder at the screen. It shows a painting from medieval Italy, a single wicker hive drawn as a cross-section. The walls are thick and smoothed like gathered skeins of golden thread; the comb inside is depicted as a series of horizontal levels like the floors in some grand building, with sculpted white pillars holding each one very neatly in place.

It's an artist's impression, and parts of it are accurate (the dark colour of the comb is just how I remember it); but it is more human than bee. The comb wouldn't have formed in horizontal layers like this, but vertically – and there would have been no need for pillars. Perhaps the artist split open a dead or abandoned hive and, finding a structure inside, opted to present it as a human society in miniature. Because, right up until the beginning of the Enlightenment, there were no lids – the inside of the hive was sealed from view. Early beekeeping was mostly just *keeping an eye on*.

'What's this?' Becky points to a piece of paper.

'Oh – it's a list of old words for honey.'

'*Milit, mez, mesi,*' she reads. '*Mit, mitsu, mi, mil, miele, mel.*'

They sound good read out loud.

Preston describes how the words for honey share a single root across Indo-European languages, tracing all those parts of the world where the western honeybee *Apis mellifera* evolved and spread. (It is thought that at some point languages along the Germanic branch made a split, and began to describe honey by its colour – from which came the Old Norse *hunung*, the Old German *honang*, the Old English *hunig*. Which is how we arrived at *honey*.) The words for *bee* are less similar; the Aryan and Germanic *bai* and *beo* are unrelated to the Greek *api* – maybe because for the first beekeepers the focus was the honey, and not the bees themselves.

Which reminds me – I am supposed to be choosing a hive, not wandering around through ancient history. I close the dozen-plus tabs.

'Which kind will you get?' Becky says, pulling on her coat as she makes for the door.

'I don't know,' I say, turning back to the screen. 'There's so much *choice*.'

I don't get a skep. That afternoon a one-line email arrives from Luke with a web link copied and pasted. *This is the one to go for.* And in the end I do go for it, a top-bar hive – more like a log on its side or a boat with four feet than anything resembling a house-shape.

Under the roof is a rectangular cavity that tapers towards the bottom, with a series of removable wooden bars rested widthways along the top, from which the bees build their comb freeform. Top-bar hives are completely frameless – the colony is given no further direction than a thin wax strip running lengthwise along each bar. They're cheaper than conventional hives, and relatively easy to make. The website describes them as offering a more *natural approach*.

I like the idea of a *natural* hive, but when I think back over the early hive forms I'm not sure I could say which were the more natural ones, or where the naturalness might be said to have started or finished. That thought of framelessness is intriguing, though. I like the idea of leaving the bees to build from scratch; of letting them follow their own forms. I've been so concerned with what I'll do without Luke around to guide me, I haven't stopped to consider that I might actually be guided by the bees themselves.

The website shows photos of the top-bar hive from various angles, but the background is white. Whatever landscape there was has been removed, so that the hive is without context, floating like the house in Corinne's drawing. Because of this I struggle to picture where it's coming from, or what it might look like once it's here.

I order from Elvin who lives in Kent and builds hives in his shed. I call him later that week. I'm sitting on a park wall beside Ellie, our legs dangling over a rose bush – we met up after work to get some food.

He'll courier it over when it's done, Elvin says. And then I just have to be in to meet it. I have some questions for him, since I'm still trying to filter the various warnings and advice I've been given.

'I've heard bees swarm more in top-bar hives,' I say. 'Is that true? What do you do to stop them swarming?' He begins to explain, something about rearranging the bars, leaving space near the entrance to avoid overcrowding inside – but the connection is bad, his words skip, there's buzzing and then the line goes dead.

'Everything okay?' Ellie passes me a plastic takeaway tub and a set of chopsticks wrapped in a paper packet.

'I think so,' I say, nearly scalding myself as I lift a corner of the lid and a shoot of hot steam escapes. 'I couldn't hear him very clearly.'

'*Hive*,' Ellie says, 'it's a lovely verb. A funny mixture of pulls and momentums.' Ellie works for the Oxford English Dictionary. Her job is to update and revise definitions, one word at a time. Sometimes a word takes weeks.

'What are you working on at the moment?'

'*Sky*,' she says. 'I finished *Cheese* last month. Yeah, *Hive* is *good*. It's got all those associations around gathering and coming together, storing things up. But then there's also hiving *off*; breaking away, making separate. Hey, maybe you've been doing some hiving away recently. I've hardly seen you since Christmas, you've been so ensconced in all those bee books.'

'I guess it's all about to get a lot more concrete,' I say. 'Soon there'll be a real hive in front of me. Noun form.'

The holiday season is over, and I'm back to work again – determined to do things better this time around. I'll adapt, I tell myself. I'll work faster.

But when I arrive there's a new set of 'key performance indicators' on my desk – a New Year's greeting from Head Office. Thirty fresh targets, divided into six neat categories, which we are now required to report on each quarter. Everyone's fuming about it, thronging the corridors, pressing noisily into the staff kitchen. I want to throw the performance indicators in the dustbin.

*

Honeybees were one of our first domestic species, over six thousand years ago – but what is surprising when you stop and think is that they've never been fully domesticated. When a swarm of bees leaves a hive today they're as wild as they ever were – not reliant on the hives we've made, and just as capable six thousand years on of following their own instincts about how to live.

I can't help feeling it is some great victory on the part of the bees that they've managed to retain their independence despite all those centuries of being kept. But it's created a tension in human societies, where people want to be able to own things and sort them clearly into category and type. In *The Hive* by Bee Wilson, I learn how in ancient Rome there were two categories of animal: wild ones, which could be possessed temporarily, and domestic ones, which could be owned. Honeybees, which could be kept but not tamed, didn't fit into either. In fact, the Roman encyclopaedist Pliny the Elder argued that, *not either tame nor wild*, they occupied a middle, indeterminate category all of their own.

I'm feeling on-the-fence about domesticity myself, I think, pushing open the front gate at the end of my first week back at work to find a fresh littering of crisp packets and beer cans behind the wall. Hannah and Darren's drainpipe has sprung a leak, and someone's dumped a mattress outside the house opposite. This area is increasingly given over to rented accommodation; you can tell by the air of undernourishment about the houses. The residents are temporary, and the landlords live miles away – there isn't a lot of motivation on anyone's part towards upkeep. In fact, this street would fit very nicely into Pliny's 'not either tame nor wild' category at the moment.

I unlock the door, step over the junk mail scattered across the mat and plod through into the kitchen. Everything is *in place* at our house (apart from the blinking heating), but somehow we haven't yet found a fit. In the evenings Becky and I have both been arriving home tired and drained and a bit senseless; ready to dump and refuel before beginning again. She goes for long walks, or out to meet friends. I take flight into books, as Hannah and Darren's TV hums through our adjoining wall.

In his studies of comparative religion, the historian Mircea Eliade observed that for traditional societies, the home was the place from which the world made sense. This was possible because it stood at the crossing point between two axes: the vertical, which is connected with the world of the gods and the world of the dead, and the horizontal, along which all journeys are made in this world. I read this in a book I picked up today by the writer Dougald Hine, *The Crossing of Two Lines*, and it made me think of the hive. Those flights running out and back from the entrance; the comb inside, building down.

In the kitchen I open the fridge and sniff a pot of week-old yoghurt, then put it back again. I'm thinking about the wildness of honeybees, their irrepressible life – that ability to build and make a home sitting right there inside them all the time. My own capacity for building and making feels pretty stunted at the moment, and in fact I wouldn't mind asking Eliade a thing or two. If home is the place from which the world *makes sense*, then what do you do when, even standing in your own house, your senses feel lost, or blocked, or broken down?

Eliade's work has also been cited by John Berger, who in his book *And Our Faces, My Heart, Brief as Photos* set about locating the origins of 'home'. It seems inseparable from houses now, and from notions of domesticity and ownership. Yet in its original meaning it referred not to a building or even a geographical location, but a state of being. A place at the *heart of the real*, according to Eliade. A place from which worlds could be founded; a place where meanings are made.

There are heart-shaped ornaments in Mary's window; I notice them there as I lock my bike to a street sign.

I've decided it's about time I got to know some other beekeepers in the area; it might be easier to step into the role of beekeeper, I've been telling myself, with a few experts around.

That old piece of folklore about bees that have been

freely given or arrived of their own accord being more likely to do well has been playing on my mind. It's tempting to think that the tradition had a practical function, ensuring the direct passage of skills from one beekeeper to another but also requiring an attentiveness, on the part of the receiver, to what makes a hive inhabitable. You can't receive a colony unless you're willing to become a part of a network; you won't catch the attention of a passing swarm without some understanding of what it is the bees might need.

I came across a website for the Oxfordshire Natural Beekeeping Group. *Natural Beekeeping* – that's a phrase that keeps popping up recently, and it marks a shift away from some of the techniques developed by modern beekeeping towards an approach that is less focused on the honey, and more on the bees. *Less intervention*, the message seems to be. It sounds good to me. But it feels like swinging from one extreme to the other to join the ONBG in the same week that I've been reading about factory-produced foundation sheets, and I wonder if they'll smell an intruder. Anyway, they're meeting tonight, and I've been sent directions to Mary's house.

Mary lives in a semi-detached house just beyond the eastern bypass. I can make it out, just about, in the window-lit dark of after-work January. There's a bright-yellow lichen on all of the trees along her road, but only where they're facing the traffic. It glows under the streetlamps like the stain on an X-ray after you've swallowed a tracer.

Her wind chimes jangle under the eaves. 'Come on in!' she says, taking my coat. Her house has a smell like nutmeg. We arrange ourselves on sofas and stools as she hands around teacups painted with tiny wildflowers. I'm introduced to Paul, who organises the meetings; Helle and her son, whose extended limbs and heavy fringe mark him out as the group's only adolescent; Jude, who arrives late and spills coffee on the sofa; and Mark, who drove twenty miles to get here. That's what he says when he arrives. 'I drove twenty miles to get here.' His head is closely shaved and he rubs a hand over it each time he speaks.

'We're low on numbers in the winter,' Paul explains, a little apologetically. 'In the summer we're a much larger group.' I look around at the tightly packed room and can't imagine how they'd fit.

Mary bustles back and forth with tea and coffee and now Paul takes a piece of honeycomb from an old shoebox, which we pass around. It's old, and dark with the repeated laying and cleaning and the build-up of debris that goes on inside the hive. It still smells of honey, too.

Next we pass round a plate of biscuits. Paul asks if I have a colony of my own, and I say that I'm awaiting delivery of a top-bar hive. 'You want to hear about Mary's top-bar!' he says, and beckons for her to sit down.

Mary retired a few years ago, and built her own hive after her husband gave her a set of power tools for Christmas. She found a video on YouTube with step-by-step instructions. 'I was in and out between the computer and the garden,' she grins. 'I'd watch a bit and pause it, then run out and add a bit more.'

Paul does a quick recap to find out whose colonies swarmed over the summer.

'I lost count,' Jude says, reaching for a biscuit. Her neck is lost somewhere underneath rolls of scarf and jumper. 'Could've been three − I saw three. But it could easily have been more.' Jude hasn't opened her hive for years. The legs are so rotten, she says, that sooner or later it's going to collapse completely − and I don't know if it's pride or bloody-mindedness in her voice.

I take a bite of biscuit, turn it around in my mouth. I'm not sure about this. Isn't beekeeping also about *up*keep? Taking care of the hive, at least, even if you're not meddling with the bees? I'm not sure I'm so interested in beekeeping if it isn't also about *opening* a hive.

Jude's approach is clearly extreme even among a group of low-intervention beekeepers, and I see Paul shuffling in his seat. There are important reasons for opening a hive, he carefully reminds us. It allows us to check for disease and make small adjustments to ensure the conditions are right for a colony to grow. Jude folds her arms and humphs. 'Well no one's going to open this one,' she says. 'The lid is jammed shut.'

That'll be the propolis, I think to myself, remembering Luke explaining the sticky resin made by bees from buds and sap to fill unwanted gaps inside the hive. It's like trying to recall a language I began learning once, but never practised. Words I've heard before but not for years have begun scudding and slipping about my skull, light and liquid and uncoupled from their meanings. I'm still fitting them together when sentences have already finished, the talk moved on elsewhere.

The room begins humming with conversation; there are always two or three happening at once. Now and then Paul interrupts and gathers us in, before sending us off again on a fresh line of discussion. There are divergences within the group. Not everyone agrees on the right way of keeping bees, and my plan to locate an expert or two who can tell me the right way of doing things seems a little naive now I'm here. They're talking about bees, but also about chemical pesticides and microscopic parasites and a new legislation that may see inspection programmes rolled out right across the UK.

I sit and listen mostly, and wonder what the old bee-keepers talked about if ever they met up in this way. The group here are comparing notes. They're sifting for clues, looking for patterns, and the patterns aren't always visible at first glance. Last summer Helle noticed the wings on some of her worker bees were looking frayed. At first she couldn't work out what it was; assumed it must be something happening to the bees in flight. In fact the cause was inside the hive, a disease known as deformed-wing virus, which affects bees in the larval stage, meaning that they are unable to fly.

Helle caught the problem quickly, and the colony recovered. But there is something sinister about not knowing where to look, what to do to make things better, and it occurs to me now, as I listen to the group speak, that the way we talk about bees and how we look at them is changing – is *having* to change – as we find ourselves in an increasingly uncertain and complex world.

Paul seems often to be bringing in outliers, softening extremes – or else playing devil's advocate where he sees too much consensus forming in either one direction or another. He's a member of the British Beekeepers Association, too – a more established and more conventional network of keepers, which has a proper membership and training programme and a shiny website.

'So he can keep an eye on the enemy!' says Jude, and laughs.

'Absolutely not,' replies Paul, without smiling.

Mark who drove twenty miles to get here has never been a beekeeper. He's just thinking about it for the first time, and has brought some photos of his garden along with a lot of questions. 'CCD,' he says, rubbing his head. 'What's CCD?'

The conversation stops, and the group leans in a little to hear Paul's reply.

I've read about colony-collapse disorder in the papers – a phrase used to describe a particular pattern of losses to bee colonies in America, which began around a decade ago. Beekeepers across several states found their colonies failing. Up to 90 per cent over a single summer, in some places.

'. . . Similar losses have been reported in Europe,' Paul's saying, 'but none have fitted the exact criteria for CCD.'

'So we're all right then, eh?!' Mark grins, and laughs, and looks around. Everyone is busy listening to Paul.

'There have been losses to bee colonies throughout history,' Paul goes on, 'usually associated with viruses affecting bee populations on a large scale. So although it's alarming, there's nothing especially unusual about a period of loss per se. Until you look at the *symptoms*,' he says, tapping his chin nervously with two forefingers. 'The symptoms of CCD point to something very different.' Mark has gone very quiet. 'Not a disease in the standard sense, since it's possible that no single cause exists. With CCD, a hive is literally *abandoned* by its workers. Beekeepers open their hives to find them emptied of adult bees. There are no dead bees inside or around the entrance, as would normally happen in the case of a parasite or disease.'

'Creepy,' says Helle, shivering her shoulders.

'They leave?' says Mark.

'*Vamoose*,' says Jude.

As though the hive stops making sense, I think, wondering what happens to make them quit. A build-up of stress or circumstance collects and heaps up inside the hive, until – in a complete reversal of their natural drive to produce food and raise brood, and despite the fact that there are scant habitats available for them to go *to* – they reach a tipping point. Exercising their only remaining source of agency, the right to leave, the workers desert their colonies, leaving behind the queen and brood who, without care, will die within a few days.

There's a pause in which Mary gets up to fetch more tea, and Jude reaches for another biscuit. And then Mary arrives back and suggests that Mark shows everyone the pictures of his garden, which he does, after a bit of persuading. And everyone looks and nods and reassures him that yes it is a fine place to house a beehive.

*

I like the thought of a low-intervention approach to beekeeping, but when I'm in the garden a few days later I realise I'm not sure how far back to step. How much room to give a colony and still learn something about who they are and what their needs and fluctuating rhythms might be? Both seem important. To step back, intervene less; but also to get to *know*, to rebuild or re-form a relationship with the natural world. I have a sense that I might learn something from this creature that hives and builds and journeys ceaselessly.

With the early hives, intervention wasn't an option – so I wonder what the old beekeepers gleaned about the life of the colony; how far it's possible to understand the bees without lifting the lid to look.

There must have been glimpses of the hive's interior. As they reached in, pulling the comb loose, the beekeepers must have noticed the toughness of it; the way it curved and bloomed. The sameness of the cells, and the way the colours drifted from white to gold to dark brown, like the tide marks on a beach. Since there was honey in some cells, but not all of them, the beekeepers must have wondered what else was in there. And how did it all get made from what went in – those strange yellow orbs that the bees carried, seemingly, with their feet.

They are stones in their feet, Virgil said. Weighting their legs like a plumb line, and offering stability in flight. Honey comes from heaven, according to Aristotle. It is *the saliva of stars*, said Pliny – or else *the perspiration of the sky* or some kind of moisture produced by *the air purging itself.* Either way, so he thought, honey falls to earth from a great height – which accounts for its varying colours and consistencies across different sites and seasons; it picks up *a great deal of dirt* on the way down, sliding over foliage and becoming *tainted by the juice of flowers, stained with vapour of the earth* and later *soaked in the corruptions of the belly* as the bees ingest and carry it back to the hive.

This reads at first as a kind of fantasy – the curiosities of the great classical thinkers given free rein to wander. The 'stones' in bees' feet were in fact pollen inside pollen baskets – the small hairlike grooves on the insides of their knees, where that yellow and sifting dust from the inside of flowers is brushed and licked into a wad firm enough to carry. When a creature is as strange as a honeybee, you have to stretch a long way in your mind to reach it. I think of Pliny's imagination doing a loop-the-loop when he describes honey as star saliva.

But what is noticeable about these ancient sense-makings is how – just occasionally – they seem to possess a kind of insight. Bees are deaf, wrote Aristotle. And so it was found, under a microscope many centuries later, that they lack an auditory system. In *Bee*, Claire Preston describes a common belief that bees were terrified of thunder and lightning, noting that it's true they can sense when a storm is coming – although whether they are afraid is anyone's guess. Perhaps they are excited.

It's still winter, and that space by the far fence is as bare as it was the day that Jack and I took to it with pruning shears. Maybe there are a few more dead leaves blown up against its sides. Spring feels a long way off, and despite the hive tool and the bee brush having arrived through the letterbox, I am feeling no better equipped to welcome a whole great massing colony of bees into the garden, or into my care.

There's a robin in the garden, poking about in the soil. I hover on the threshold, hoping he won't notice me or that if he does I won't disturb him. Suddenly Hannah's cat emerges from under the hedge, and there's a flash of wing and feather as it makes a dive for the robin and the robin scarpers.

Cracks of time open at either end of my day. When I wake and before Becky gets up I pull out books, turn pages. At the library I make reservations, add my name to waiting lists. I find I want to trace the development of the modern hive; if I can map out the journey that led to the birth of modern beekeeping, perhaps I'll have a clearer sense of how to keep the bees from here on.

By now four books in particular have become mainstays: Claire Preston's *Bee*, Bee Wilson's *The Hive*, Hattie Ellis's *Sweetness & Light* and Eva Crane's *The World History of Beekeeping and Honey Hunting*. They sit stacked up beside my bed, a flotilla of females – writers who voyaged deep into all those centuries of beekeeping history, much of it male, and navigated a course through. I underline passages, flit between chapters. *The Hive* is soon marked with a coffee stain.

Tentatively, I begin navigating my own course through. I find I'm tracing the development of the modern beekeeper, as well as the modern hive. I want to know where our involvement in the hive's inner processes began; where a curiosity about the bees became an impetus to see and begin controlling what happens inside.

'You're a beekeeper?' asks the librarian, eyeing my oversized pannier bag as I heap in a fresh stock of books.

'I was,' I say. 'I mean – I will be.'

One evening, in Eva Crane's book, I read a curious story about Aristotle and his bees. Apparently he built a window, not for looking out but for seeing *into* his beehive. Thick walls, a thin crack down one side and a transparency inserted. The window didn't work. According to the story, the bees obscured the glass with clay. It would have been propolis, not clay, but still. I read it as a call from the ancient past urging caution: don't let intervention run unchecked. To look inside was to disturb the proper functioning of the colony – a disturbance that could be remedied from inside, by the bees, who appear capable of accepting or rejecting an intrusion at will.

I'm intrigued. I begin looking out for more of these openings; cracks made in the hive walls. I learn that Pliny described hives made from transparent *mirror-stone*. It was probably the mineral mica, and is unlikely to have made a whole hive, but – fingernail-sized, ear-shaped, eyeball-like – it seems that windows did appear throughout the history of beekeeping, and there were people who caught brief glimpses of the hive's insides, before the bees covered them over.

I've reached as far as the seventeenth century, and it's getting late. I have work in the morning, and really I should be getting to bed – reserving my hours of wakefulness for the present day, not journeying down through the past. But I've got to the point where the age of scientific discovery begins, and a fresh focus develops in relation to honeybees. Distinct from the practical concerns of common beekeepers, the bees *themselves* have become an object of study.

Through the Middle Ages, beekeeping had been the work of the lower classes, and hives were kept solely for the harvesting of honey. But with the growth of the natural sciences it drew the attention once again of learned scholars, and a new generation of apiarists rose up, many of them clergymen – educated men with time on their hands and a keen interest in the wonders of the natural world. These early scientists of the Enlightenment era began to observe honeybees meticulously, methodically; and before long they began to question, then to disprove, many of the accepted ways of thinking about the hive.

Hattie Ellis describes how The Reverend Charles Butler, great-great-grandfather of the English natural-ist Gilbert White, was among the first to challenge the received wisdom of classical authors. He came to believe that drones were a part of the reproductive function of the colony, although he couldn't say how. He noted also a peculiar activity among bees returning to the hive. They displayed a soft *shivering*, he wrote, as they waggled at him.

I stand up and stretch. Ball my fists, rub my neck, remember the pile of paperwork I will have to return to in the morning.

Moving over to the window I touch the curtain to make a gap for seeing. Outside it's dark apart from the light of the street lamps, and of TV screens flickering in some windows. A double-decker passes and the floor trembles a bit. The bus is empty except for the driver, its insides brightly lit. They're tinted a whitish blue.

I think about those early scientists working amid all that excitement and change. There must have been a parallel search happening alongside their explicit investigations. They needed a new *language* to accompany this fresh approach to seeing and thinking about the world. How to describe something impassively, straightforwardly, without recourse to accepted theory? How to balance a sense of wonder about the natural world with the need for rational objectivity? How to find words for what you see, and – more than that – to deduce meaning from it; to explain it?

As I climb into bed my phone beeps – a message from Elvin, who seems to do most of his hive-building at night. *Should be ready in a few weeks*, it says. *Do you want the roof tin-plated?*

*Yes please*, I reply. It is good to be watertight.

That night I dream I'm standing inside a closed room. The wallpaper is the same kind I had in my bedroom as a child, with woodchips sealed in between the paper layers so that the whole wall bumps and mottles like there might be a thousand insect bodies trapped inside.

I am holding a cardboard box. Inside the box is a colony of bees; I can feel them moving around inside. I want to open a window; the window is locked. A hole opens in the base as they begin chewing the cardboard sides. They're huge. They have teeth.

The following day I shoot out of work on my lunch break, hot from the pent-up atmosphere inside and desperate for an escape route. It's been tense all morning. Everyone chasing after deadlines, sticky with each other and full of angst. Me too. I feel like an out-of-place and poorly tessellated shape, full of gaps and overlaps and awhirr with unwanted frictions.

I begin stamping out the frictions as I walk, as though there might be answers inside the pavement. All those targets, that push for productivity, the paperwork and increasingly standardised procedures; they might be pulling us into line, but they don't feel very – I round a corner, search for the word – *human*, to me. They don't cultivate trust, or care, and they aren't where ideas are made. Those individual desk cells leave us isolated – and not in a way that leads to any kind of self-realisation, either.

There are no answers in the pavement; there is just more resistance and silence.

As I reach the high street I'm unsure which way to go; how to break free of this stuck-inside feeling. Then I remember the observation hive. Paul mentioned it at the ONBG meeting. It's installed at the natural history museum not far from here, and I think I can make it there and back on a lunch break. I check my watch; I'll have to be quick. I am already half-running down the street.

The sun is out and the wind is sharp. I hurry across wide-open squares and take shortcuts down narrow side-alleys, squinting up at the great grand buildings and stepping through wall, roof and chimney-top shapes as the sun makes shadows of it all. The sun is achingly bright. I zip one way and then the other, losing a sense of which side the sun will come from next, and sometimes anyway it comes reflected: around a corner between two crumbling walls it leaps off a basement window. My hair whips, I gasp great gulps of air. Everything is fresh and tossed around.

Inside the museum entrance is a vast stone atrium. A gargoyle-encrusted ceiling soars high above my head. The whole place echoes with the voices of a busy school party, and with their feet on the flagstones. There's a shop, selling museum-logo-embossed pencil sharpeners and plastic animals, on one side, and an unmanned information point on the other. A sign in front of me points visitors Left, Right, Up. Toilets, Cafe, Beehive.

I go right, and up. Climb a stone staircase with high windows separated into a hundred panes. You can't see anything but light through them, and the light floods the stone steps. My footsteps echo like the school children's voices, but now the school children are behind me. Like comb, I think, looking at the panes. The staircase touches the window, and then it curves.

At the top is a room with dark-green carpet and glass cabinets lining the walls, and away from the stone and the busyness downstairs there is a sudden hush. I'm the only person up here; the observation hive is obviously not drawing a lot of attention at this time of year. Inside the cabinets, plastic plants and taxidermied animals are arranged in woodland and desert scenes – they're all looking a little tight for space. I peer closer at a wild boar but can't decide if he's yawning or baring his teeth through the glass. The fur on his shoulders has a greenish tint; there must be a filter fitted over the light bulb. I turn round and spot the observation hive standing over by the window.

The Age of Scientific Discovery brought with it a fresh impetus to gain access to the interior life of the colony. In 1649, The Reverend William Mewe, inspired by Pliny's transparent hives, constructed an octagonal hive with multiple tiers, each one fitted with a small window sealed by a hinged shutter. *I can take a strict account of their work*, wrote Mewe, *and thereby guesse how the rest prosper.* With developments in manufacturing it became possible to produce glass in sizeable sheets, and what we now call the 'observation hive' arrived – a single piece of comb sandwiched between two panes, with a whole colony on display. *Not a single bee could escape my notice*, wrote another scientist-cleric, The Reverend William Dunbar, in 1820.

As the scientist-beekeepers exercised their eyes and rational-thinking minds, fresh discoveries came quickly. But there was something else, too. Through the glass, something more subtle and less observable was happening between human and honeybee.

Reading their accounts, I've been struck by the attitude among these early scientists of increasing scrutiny and surveillance. As though, along with new capacities for seeing and knowing, came intimations of control. *There was not a cell . . . in which we could not examine what took place at any moment*, wrote the Swiss natural historian François Huber in the 1790s. *I might almost say that there was not a single bee with which we did not get personally acquainted.*

I scoffed at that word, *acquainted*. Some chance, I thought, thinking back to Luke's hives and remembering the difficulty of picking out individuals from that mass of a seething colony.

The museum's observation hive is raised up on a wooden platform. There's a lever to pull, and the whole thing rotates for you. A piece of plastic tubing runs from the base of the hive to a hole in the wall and outside, but it's empty today; too cold for flying. I clamber up and put my face to the window. The comb inside is clustered with bees, which are also clustering at the glass. Close up, I can see marks like backbones on the undersides of their bellies, except that honeybees don't have bones. They have exoskeletons, like armour plating; their insides are soft and lack any internal hard structure.

Gathered together like this their wings could be cobwebs, if you weren't focusing properly. Chitin, it's called. The same stuff their exoskeletons are made of, a substance a bit like our fingernails. The exoskeletons are rigid and watertight. When I learned this, I assumed it was to stop the rain getting in. In fact it's about retaining body moisture; it stops them drying out.

I move closer until my nose is nearly touching the glass, and the bees bristle and stir inside. A CCTV camera turns its head in the corner, and a red light winks at me.

In the writings of those beekeeper-scientists, I think I've happened upon a tipping point in the history of the hive. Here was one instance where the motivation to learn – to follow a curiosity, and find out, and understand – overlapped with that other, unshakable desire to get involved. To approach and intervene, even to gain control. What happened next would be the direct result of that overlap and those feelings of possibility.

It is also an instance of how the relationship between beekeeper and bee can become a little tangled. In one letter, William Mewe wrote to fellow beekeeper Samuel Hartlib of a curious effect he'd encountered among his bees. They were, he claimed, changed by his looking. Since he began regular observations of his colony, honey production had *doubled*. Something was happening, he believed, between beekeeper and bee. The whole colony was modifying itself under, and in response to, his gaze. At no point in his letter does Mewe mention whether the effect ran both ways, so that he was also changed.

I'm wondering how this colony of museum bees might have modified itself under and in response to the gaze of the taxidermied woodland and desert species, when there's a shout behind me and a kid runs into the room with a blue plastic sword. He makes a warrior pose at the wild boar and I giggle and he jumps, then turns and does a warrior pose at me. I grab my neck with both hands and make a noise like a death rattle, and his eyes light up like he can't believe his luck.

'*Mikey?*' We hear his mum coming up the stairs. '*Mikey!*' He pauses for a moment like he's unsure whose world I belong to, the grown-ups' or his own. And then he's gone.

I step down from the platform, head spinning, remembering lunch. Nothing about this hive feels familiar, I realise. Seeing the colony on display like this feels very remote from opening the hives with Luke and, despite the proximity, I feel strangely numb. Everything is visible, and readied for inspection; all other senses are turned to mute. I can't smell, hear, touch anything from behind an observation pane; and the bees can't smell, touch, or hear me. Like this the colony seems oddly stilled, more museum specimen than living thing. And I want to get away.

# February

It is all very well to go delving into the past, but if it's all you're doing, it can begin to look like an avoidance of what's actually happening *right now*. One thing happening right now is that on a farm near Banbury a colony of bees is clustering, waiting for the spring. Another thing is that somewhere in Essex a cedar plank is being sliced and planed and drilled and sanded to make the sides and base of a top-bar hive.

If I'm honest, I've been trying not to think about it. Just at the moment the thought of a real colony arriving is like too much added load. I've been keeping it together at work, keeping myself in place, but I feel brittle and stretched and it is taking all of my energy just to stick to type. The last thing I need is another difficult-to-define creature on my hands; another responsibility.

It would be much simpler to stick to reading about bees in books, where the words don't move around like real bees do and everything has been laid out neatly in order. Becky suggested yesterday that I should do a bit of reading on how to keep bees, present tense, since the bees are actually coming. Probably she is right. But I'm not ready yet. Avoidance tactic it may be, but for now I am following the history.

I'm clearing out cupboards in the kitchen one afternoon when I get a call from Luke. He's in Soho, and about to check on a hive – I picture the beekeeping suit and the box of frames he'll have strapped to the back of his moped.

'How are you doing?' he says.

'Good!' I tell him about all the books I've been reading. 'I'm really getting into it,' I say, throwing the names of a few centuries-old beekeepers at him, just to see if he knows any, and in fact he knows all of them. 'But – I'm not really sure what I'm doing. I mean – why read about the history of the hive at all? I'm not sure what it is I'm trying to find out.'

'Well I wouldn't bother about that,' he laughs, and I hear the sound of a car horn in the background. 'What's the point of finding something out if you already know what it is? Maybe there are some questions you're trying to ask yourself,' he adds. 'You just haven't found the right words yet.'

There are no questions I'm trying to ask myself, I nearly say; I am just interested in the hive. But that wouldn't be quite true.

'I'm worried I won't know what to *do*,' I hear myself telling him. 'How to look after a colony, I mean. We've been building hives for thousands of years, but I'm not sure we've ever really learned how to *keep*.' In fact, all our self-assuredness about being able to secure and manage our environments is looking decidedly strained, I want to add – but I can't get the words out. 'I thought if I could get a handle on the hive itself, how it's changed over time, I might get a better sense of how to keep it.' And I tell him about the Enlightenment scientists, all excited and caught up in their own powers of observation.

'Hm,' Luke says, and I wonder if he's followed me or if I lost him somewhere around the 'thousands of years' bit. 'Well, if you really want to get into it, François Huber would be a good lead to follow,' he says, and I remember the Swiss natural historian and his claims about being acquainted with the whole of his colony. 'He represents a bit of a turning point, when science began to intersect with everyday beekeeping. He tried to *apply* science, make it useful to beekeepers in a practical way.' He pauses as a siren passes. 'Anyway,' he says. 'What about your hive? The real one, I mean.'

'It's nearly ready. Should be here in a couple of weeks.'

'Good. Won't be long now. Soon the bees will be there too, and then you won't have time for all this reading.' Which panics me slightly. There is still a lot more to know.

Luke is good at making you believe you can do something even if there are a lot of voices in your head arguing to the contrary. He doesn't try to understand everything; he says he's still learning all the time, and in fact the learning is all part of it. Sometimes the novice beekeepers are even the better ones, he told me once – they have less of a picture of how the hive *should* be, so they pay more attention to what's actually happening when they look.


In 1792 Huber laid out a record of his investigations into honeybees in a series of letters to the acclaimed naturalist, his hero Charles Bonnet. And now – over weekends, between work and life and seeing people – I begin making trips to London, slipping into the British Library to read them.

By the time of Huber's writing, scientists had come to understand the true materials from which honey and wax are made, and some of the processes involved in their production. The colony is ruled by a queen and the celibate workers are female. There are also male drones, larger than the workers and darker in colour, with bigger eyes that almost meet in the middle, so that there is no room left for a forehead. The new knowledge had transformed scientific understanding but bore little relation to the practical stuff of beekeeping itself, and it was to this that Huber now turned his attention. He wrote to Bonnet of his concerns over traditional practices. It didn't make sense to him that beekeepers should destroy a whole colony at harvest. Thousands of years had passed since people began keeping bees, but in all that time the hive had remained much the same. Most beekeepers still used skeps, and employed practices that had been in place for centuries – but these were beginning to look nonsensical. The landscape and agriculture had changed. There was less forest and fewer wild colonies around; the bees were not so easy to replace, yet still the beekeepers killed them.

I can think of a few parallels between Huber's world and my own. That feeling of the world as a place of uncertain resources; of living-practices being out of kilter with them. Huber didn't think much of the observation hive. Bees weren't made to live on a single comb; *they are taught by nature to build parallel combs . . . a law from which they never derogate [unless] constrained.* Along with the letters, he sent Bonnet a collection of diagrams and instructions for the 'leaf hive', his own design. Twelve identical wooden frames, each one hinged to the next and all tied together with string. It looked like a box, at first, and then like a book, the frames like pages as they opened out. I imagine the combs like crusted pages, crammed to their edges with bees. The frames must have stuck, a bit, with propolis, which would have made a cracking sound. And then the bees, shifting, hiding, shuffling themselves.

With Huber's design the beekeeper could observe the wooden frames one by one, thereby gaining a sense of the hive as a whole. There is capacity, he asserted, to make small adjustments to promote the healthy functioning of the colony. The keeper could separate the frames. Remove pieces of honeycomb and close the hive back up, leaving the bees to reconstruct the missing parts. He was making a case for a new and more efficient hive design, but he was seeking to reset the terms of a relationship too. There is room for manoeuvre, he seems to have been saying. It is possible to make alterations to the hive itself, but beekeepers need to change their own practices too.

*When we have all the combs before us . . . we see how abundant the provisions are and what share of them we may take away.* Huber wanted beekeepers to see into the hive, but was wary of intervening too heavily. He urged moderation in the harvesting of honey and wax, and argued that *to compensate for this moderation every means must be employed to promote the multiplication of bees.* We should take a little; and do everything we can outside the hive to help them thrive.

I'm surprised to find that his letters – while scientific and rigorous and occasionally pedantic – are alive to touch and sensory detail. The hive opened and the comb sat exposed. With regular inspections the bees became *tractable* (easier to work with), Huber noted, careful not to suggest that he had tamed them. He wrote also of changes in the light: the shock of sunlight when the hive was opened seemed to excite their fear; the bees shrank into the comb, concealed themselves. They were *less tractable in the night or after sunset.* Sitting in the library, the wintry sun peeling in through the skylights as I feel my shoulders loosening and a knot unfurling, I begin warming to him.

The bees' world was so alien – it followed laws and logic that bore so little relation to Huber's own – that, aided by his assistant François Burnens, he was obliged to proceed without a map, carrying only his own powers of deduction with him. This led him to some curious conclusions. *The air which we exhale appears to anger them*, he wrote, and experimented with bellows instead, which scattered them.

Then one day Huber noticed something truly surprising. The combs were forming just the same as they always had, vertically and parallel to one another, but now he found his attention drawn to the spacing *between* them.

He measured it. Checked himself, and measured again. And perhaps the hive reversed itself in his mind, so that he saw its negative space. *This leads me to what I believe is a new observation*, he later wrote. Throughout the hive the spaces between each comb were uniform, and this, he became convinced, was essential to the health of the whole colony: too far apart, and the bees would be less able to capture and preserve heat; too close, and they would be unable to move around freely.

What I don't learn until this point in reading the letters is that Huber was almost completely blind.

I sit back.

He's *blind*?

This man who claimed to have been particularly acquainted with his bees, and who was so passionately committed to observation? What to make of that, if seeing was the one thing he couldn't do? All that time I've spent on the trail of the old beekeepers as they formed windows, made cracks in the walls of hives; and now here is Huber who got rid of windows altogether – who *opened up* the hive – but couldn't see.

I'm curious but also a bit wary of this man François Huber, and unsure that this lead Luke has given me is really heading anywhere useful after all. But anyway I keep on with the history of the hive because I haven't reached the modern hive yet, and without knowing that part of the story I won't be able to join the dots – won't properly understand where we are now, what's happened to the bees, or what it means to be a beekeeper today.

The next part of the story concerns another man, a pastor named Lorenzo Lorraine Langstroth who lived in the nineteenth century, a long way from Huber, on the other side of the Atlantic Ocean.

Huber had doubted that the Leaf Hive would ever become popular, and ultimately he was right. Although used occasionally for scientific study, it proved impractical for common use. The frames stuck. It was bulky, and difficult to build. But over half a century later his insight regarding the evenly spaced combs resurfaced when a copy of his letters to Bonnet fell into the hands of Lorenzo Langstroth in America; and it was there that the modern hive finally came into being.

As a young man, Langstroth had trained as a Congregational pastor, but this vocation had not been an outright success. He suffered bouts of debilitating depression and anxiety, experiencing a muteness so pronounced that he became unable to give sermons, sometimes for months at a time. Having moved from one failed church posting to another, in 1848 he resigned as a pastor for the final time. Rekindling a childhood fascination with insects, he set up an apiary in the garden. He'd come across copies of Huber's letters a few years earlier, and now he began studying these. He began to experiment with Huber's hive design and another hive developed by the apiarist Edward Bevan. *I became thoroughly convinced*, he later wrote, *that no hives were fit to be used, unless they furnished uncommon protection against extremes of heat and more especially of cold.* He doubled the layers of the hive's exterior. Next he started on the hive's insides, determined to promote a greater and more reliable honey flow.

In the Bevan hive the honeycomb was built on removable bars, but when Langstroth tried these he found the bees soon fused them to the hive walls. He tugged the comb, and the bees flustered around him. The comb was stuck. How to make it movable, he wondered, and found himself turning again to Huber's gaps.

He took this gap, the distance between combs, and began testing it in other places. *Dead air space*, he called it: a distance of three-eighths of an inch that the bees would leave open throughout the hive. Any smaller and they plugged it with propolis; any larger and they bridged it with honeycomb. *Bee space*, it later became called, since in fact it was far from dead. Langstroth diligently entered the measurement into his hive design; a series of frames surrounded on all sides by this space. Tested, it worked. The frames remained movable. He lifted them out, and put them in. Shuffled, expanded and rearranged them, as the bees shuffled and rearranged themselves around him.

Encouraged by friends to publish his findings, Langstroth patented his design and moved away from his family to live with his sister in Greenfield, Massachusetts. There he installed an apiary in her garden, and – moving between bees and words, hive and study – began to write. His handwriting was illegible. The words jumbled and knotted across the pages, unreadable by anyone except his wife. And she was miles away. He began posting her packages – a gift, his scrawl – which she carefully transcribed for the printers. *Langstroth on the Hive and the Honeybee: A Beekeeper's Manual* was published in 1853, and the hive was a success, heralding the biggest shift in beekeeping since ancient times.

Langstroth did not become rich. Despite its patent the hive was easily replicated, and anyway it emerged that a beekeeper in Poland, Johann Dzierzon, had also been working with the bee space. The Langstroth hive was reproduced and reinterpreted across continents, spawning the WBC (a hive named after its inventor William Broughton Carr), the National and the Dadant (after its inventor Charles Dadant). In place of roundedness came squares and rectangles, a set of parts that could be easily replicated and assembled. Here is the hive we recognise, the sloping roof and weatherboards and the narrow entrance like a doorway at the base. With the removable lid and frames there is little of the hive's interior now that remains unknown to us; we can say that we *understand* it. More than that, we've learned to believe we can control – even have mastery over – what happens inside it.

Back at home in Oxford I've been out in the garden, pacing the distance between the back door and the fence. Becky asks, will the bees come into the house, and I say I don't know, I think that they might. If we leave the windows and doors open then they might, maybe.

Indoors I spread out on the sofa with half a dozen books and my laptop in front of me, a muddle of text and pictures.

The modern hive had allowed beekeepers to manage disease, even to work with the colony – making small adjustments, as Huber had predicted, to ensure the conditions were right for it to thrive. Commercial beekeeping was possible for the first time, and with the rise of commercial practices, the focus among beekeepers shifted again to the process of honey-making that goes on inside the hive.

Honey doesn't come from heaven as Aristotle suggested but from nectar, which is contained within the glands of plants – a thin and easily spoiled liquid that is converted by the worker bees into a stable, highly concentrated and high-energy food source. A bee extracts nectar from a flower by poking and sucking with the tip of her proboscis. A proboscis is a bit like a curling tongue, I learn, and the nectar is taken inside her honey sac, which is like a temporary stomach. Here it mixes with enzymes as she moves around until, on her return to the hive, a worker 'begs' the nectar by drumming on her antennae. This prompts her to regurgitate it, passing it to the worker, who presses it into the underside of her own proboscis, drawing some of the moisture off before passing it on. Like this the nectar is passed around the hive from one bee to the next in a process of communal digestion known as *trophallaxis*. With each transfer, a little more moisture is extracted, and the humidity inside the hive increases – the workers have to fan their wings to keep it ventilated. When the nectar has been converted into a supersaturated solution it is ready to be stored inside the comb cells and sealed with a thin layer of wax, until such time as it is needed. At this point it becomes recognisable to us and also harvestable as honeycomb.

It's a complex process, carefully regulated by the colony and highly sensitive to outside changes – an especially cold spring or a drop in forageable flowers can drastically alter the quantity of honey produced from one year to the next. Yet over the last century, the pressure on commercial beekeepers – as on the owners of the factories and farms that have also undergone a rapid process of intensification – has been to drive up efficiency and increase levels of production. The biggest bee farms are now thousands of hives strong, and operate on an industrial scale.

The hive has changed shape again, but this time the change is a perceptual one. Where we see honey not as an active substance and part of a living system but an isolated product like a car or a cake or a cooking pot, the hive comes to resemble a factory; and that delicate and contingent process of making honey is reduced to just another point on a production line that begins in the fields and ends on the supermarket shelf. This changes what happens as we encounter the hive; it alters what we do. The job of the beekeeper-overseer under these circumstances is to iron out the creases, fill in the gaps. Remove any processes superfluous to the *real* work of making honey.

Common practices now employed by large-scale commercial beekeepers include artificial feeding and insemination, the use of antibiotics, the transportation of live bees across vast distances, and – in a chilling recurrence of one of the earliest and most harmful beekeeping practices – the culling of hives in winter. Not, this time, because no other option is available; but because it can be cheaper to kill a colony after harvest and buy fresh stock in the spring than to feed it through the cold season.

UK supermarkets are among those organisations that have led enthusiastic campaigns aimed at raising awareness about struggling honeybee populations and the importance of expanding pollinator-friendly habitats; but none have addressed the issue of beekeeping husbandry, nor the part the honey industry may have played in the population declines. In fact, honeybees are notably absent from their published policies on animal welfare.

I sit back. I'm nearing the end of this journey into the history of the hive, and as I pull myself back to the present day it strikes me that it is the bees who are the ones intruding here and not the other way around. Present by their absence, inching in like an open question to the places where they've been omitted, they've been circling the margins as I've been reading, nudging my attention back towards them.

It is probably just as well, since February is almost at a close, and it won't be long now before the real bees are arriving here. Soon there will be no avoiding them.

Even away from the books, I find parts of the hive's history recurring in my mind, like when a tender spot on your skin prickles and keeps prickling even after the touch has finished. I think of that claim of William Mewe's that the bees were changed by his looking. I'm not sure I believe that a colony of bees could truly modify its behaviour just by dint of an attentive gaze. But I do know that how we view the world has consequences, and that just by changing a filter or focusing differently we can create blind spots or open new windows, even without noticing as we do.

It's Saturday and I'm in the office, the week's work having spilled over into the weekend. I'm searching through folders for a misplaced risk assessment when a friend sends me a link to an article about a new invention, the flow hive. It removes the need for comb-building by supplying ready-made plastic cells, laid out with perfect uniformity. So all you need to do is put the bees in, and then the bees make honey. There's a tap to turn, and the honey pours out when you need it.

And just like that, we stop looking. Caught as we are in our own containers – our work schedules and production targets and all the other apparatus of our own making – the particular rhythms and processes of the colony become peripheral; secondary, at best. Except, of course, that not everything turns out as we might expect. Sometimes the things we thought were masterable turn out to have a will and a life of their own.

I'm halfway through the article about the flow hive

when the web page freezes. I can't click onto the screen or out of it. I press *refresh*, try reopening the page. A message pops up to say there's a server error; the internet connection is down. I phone down to reception, and they call up the IT guys – but the IT guys are off at the weekend.

I'm forced out. I cycle home slowly. The roads are quiet, and the pavements are full of people. The sun is out and it's making the car wing-mirrors wink. And now I notice buds have appeared on the trees; they're hard and shiny and too well-sealed to give much of themselves away just yet, but still: they're a sure sign that spring is on its way.

In the garden the space by the fence is in full sunlight. The ground has hardened since we cleared it, but there are spikes in the earth where grass shoots are beginning to show through. I poke the tip of my toe in, work a pebble free.

I know that I've begun drawing connections between the bees' world and my own. I've caught myself thinking we might be facing some of the same struggles; that perhaps we're under similar strains. It is probably not very sensible to think about them in this way. It makes everything a bit blurry, and just at the moment I want to be able to see clearly. Still, the parallels are difficult to miss, and it is not often I find shapes that fit with each other.

Back in the house there's a text message from Elvin to say the hive is ready; he's booked a courier to deliver it.

When the hive finally arrives it is in pieces. They're packed inside a cardboard box, and it rattles when it tips. The box is delivered by a man with a van who picked it up from Elvin's shed in Kent and drove it all the way here.

The first I see of it is the red *FRAGILE* sticker pasted to the front as the whole thing crashes from the van. Now it's standing upside down on the pavement.

'It's heavy,' the man shouts, and heaves it through the gate. I can only see his feet. As though he has a box for a body, or as though the box has feet. Now it crashes into a plant pot, and the plant pot rocks.

'Careful,' I say, then I pick the pot up and move it.

He sets the box in front of me. Looks it up and down. It is level with his shoulder and almost fills the doorway. It is not really a box. It's a taped-together patchwork of used boxes – printer cartridges, storage shelves, a kettle – all opened out and wrapped around each other to make one parcel big enough to hold a beehive. Underneath the *FRAGILE* stickers there are more messages scribbled in black felt-tip. *Don't drop me!* with a squiggle, and then *HANDS*. Two arrows looping and pointing.

'There's hand-holds,' he says, and shows me two hand-sized holes, one on either side. He spreads his fingers wide and bunches them, slides them in and grips. Lifts the box, then sets it down. He lifts and sets it down again.

As the van drives away I slip my own hands in and feel wooden struts inside. I don't know where to put it, the box, so I put it by the sofa. Fetch a pair of scissors, find a seam and slice it. There's a smell like the underside of parcel tape and paper fibres mixing and then suddenly the whole thing falls apart, tape ripping, the put-together structure peeling outwards as its insides tumble across the floor.

I sit down and begin picking the pieces up one by one. Here are the top bars. I run my fingers down them, feeling for the tips where I'll lay wax strips lengthwise along each one. A marker for the bees, encouraging them to build along each one separately rather than across or between, so that comb and bar can be moved and removed in one piece. It's a first bid for correspondence between beekeeper and bee, and compared with the solid wooden frames and foundation comb of a conventional hive the gesture seems hopelessly tentative; it'll be some great stroke of luck, I think, if our orientations happen to coincide.

\*

With the hive pieces piled up beside the sofa I return to the books and websites, and to the beekeepers I know. Once you've built your hive, I'm told, you should treat it on the inside and out. The outside should be painted to protect it from the rain and damp. Use a mixture of beeswax and linseed oil; or buy a can of Cuprinol. Rub the insides with beeswax, make it feel like home. Harden its outsides, soften its insides; give it a ready smell.

On Sunday, Jack arrives to help. Elvin did not send instructions, so we lay the hive pieces out across the floor and begin guessing which bits are for what thing and where. We turn things upside down and right them again. We don't talk much, and when we do, the talking is easy. Why is that hole there, look at these wine corks, the way they're whittled at one end. See here, this edge, and then a flat.

Jack asks what I've been up to recently, and I tell him about going on the trail of the beekeeping greats.

'There's something I noticed the other day. Both Langstroth and Huber had a sense missing; Huber was blind, and Langstroth sometimes became mute. Makes me wonder what that does to your experience of the world. I've heard when you lose a sense you can become especially sensitive to the others – like you pay a different kind of attention to things.'

'Yeah, except that voice isn't a sense.'

'Oh – of course,' I say, wondering how I could have got the two conflated.

'Sight, sound, taste, smell, touch,' he counts them on his fingers. 'Not voice. Voice is how you speak; it goes out. It's not about how you perceive.'

We roll the rug up, lay plastic bin bags and Sunday supplements out across the floor. I fetch brushes, and we start painting the exterior.

This month Jack's been spending his nocturnal hours on the Anglo-Saxons, and he tells me about the metrical charms they used to fix a difficult situation or a disease: 'The charms were written on objects or spoken out loud or sometimes it was about a particular action that had to be performed.'

I pick up one of the sellotaped packets, and read the hastily scribbled label. *Corks*, it says. And here's another, *Mouse guard*.

'There were spells to find lost cattle or to hurry a birth, and there was one that kept bees from swarming and leaving the hive.'

'I like the sound of that.'

'I wonder if they worked.'

We start a second coat, our hands sticky with the gummy wood preserver, which is sticking to the paint brushes too. We've been trying to fix the hive base to the frame with metal clasps, but now the base swings open.

We pick it up, begin again, our heads sitting level with the lid as we feel underneath us for the click. The sun goes down outside and we are still in the room, the sofa disappeared under cartons of wood glue and drill bits. There are scraps of scribbled paper at our feet.

3

*Bee*

# March

After all of our waiting and careful preparation the hive sits outside with a *what now?* feeling about it, an unnaturally bright near-tangerine beside all of the stripped-down greys and hard-fought greens that have made it through the winter.

In fact, winter seems to have stuck fast. That early glimpse of spring has disappeared. The winds are up, and we're still hurrying around in hats and coats. When the weather warms we can fetch the bees, but you can't go moving a colony around in the cold, so for now we rattle around the house, and we're still lighting the fire even though *forgoodnesssake* it's March already.

\*

It is around this time that I end up back at Oxford's natural history museum courtesy of Charlotte, an anthropologist-turned-zoologist whom I met a few weeks ago through a friend.

*What are you doing this afternoon?* She sends me the text message at 6 a.m. *Can you leave work a bit early? Meet me at the natural history museum? I've got us a backstage pass!*

Charlotte has red cheeks and velvet shoes and the kind of quick, alert intelligence that you can never quite keep up with – it is always spilling over into new things. When we met she'd just got back from Japan, where she'd been researching edible insects, and especially wasps and hornets. I invited her over to see the hive and she arrived on my doorstep a few days later with a bottle half-full of darkish liquid, one whole and huge dead hornet swilling around at the bottom of it.

'Hornet whisky!' she announced. 'I made it myself.'

Apparently what you do is buy a bottle of whisky, then catch a live hornet and put it inside. As the hornet drowns it releases its venom into the liquid. 'It's supposed to have loads of healing properties,' she told me. I unscrewed the cap and sniffed it.

We went out into the garden, walked up to the far fence.

'Are you ready for them?' she asked when we'd reached the hive, taking a peek under the tin-topped roof.

'Yes!' I said, infected for a moment by her air of keen proficiency. 'Well – I think so.'

Before beekeepers learned to split and breed colonies artificially, one means of populating a hive was just to leave it out like this, in the hope that a passing swarm might settle inside it. I told Charlotte this, and we stood and looked – but the thought of a swarm of bees turning up in the chilly back garden of an end-terrace seemed pretty far-fetched, to say the least.

I tried a drop of the hornet whisky while Charlotte told me about visiting black-soldier-fly farms in America. 'It could be the *solution!*' she said, and I coughed as the whisky hit my throat. 'Black soldier flies are rich in protein, and they feed off waste; they're what we'll be eating after everything else is gone.'

I told her about my trip to the museum's observation hive. 'The thing is,' I said, 'I really want to see a bee close up. A real one, I mean. Have you ever looked at one under a microscope?'

'Of course!' she said. 'Maybe I can arrange it for you – let me see what I can do.'

I'd like to know what those Enlightenment scientists saw when they leaned into the hive with that freshly rational and dispassionate gaze; the will to sort – to identify category and type. I want to lean in too. The hive has arrived, and soon the bees will be here as well – so how to approach, and begin relating to them? All together in the hive they can seem quite full of chaos, the very opposite of reason, and without Luke around I will have to learn to make sense of them on my own. So what kind of looking – what kind of scrutiny, what quality of attention – should I be aiming for? What kind of gaze is good?

I'd scoffed at François Huber's claim to have become personally acquainted with his bees, but I've been wondering all the same if it might be possible to identify individuals; to learn something about the colony as a whole by separating it out, piece by piece. A hive is not a factory and a bee is not a machine-part, and if I am to become a beekeeper proper I want to understand, as far as I can, her bodily experience – how it feels to be a bee going about in the world, with hooked knees and curling tongue, eyes and sting and armour plating.

After Charlotte left I screwed the cap back on the bottle of whisky and took it upstairs. Now it's sitting on my bedside shelf, where I can keep an eye on that huge hornet drifting around at the bottom of it.


Finally, I do manage to leave work early that afternoon and I hurry over to meet Charlotte at the museum entrance, where we wait for Polly, the museum entomologist who's agreed to show us around.

'Hi,' Polly grins, arriving in a fluster. 'Sorry to keep you waiting.' She's wearing a white lab coat over a blouse and jeans, and the kind of shoes that are made for being practical. Black, laced, rubber-soled. She looks a little out of place beside the museum shop with its bright plastic toys and visitors' maps and all the kids running around everywhere, and she quickly ushers us towards the stairs.

We turn away from the room with the observation hive this time, and enter a door marked *Private*. Here is an office corridor. The rooms are separated by partition walls; everything is a laminate beige, and there's a lot of *stuff* around. Scientific apparatus and papers are piled on every surface. Insect specimens lose their colour under UV light, so all the windows have been screened with blinds – more beige – to help preserve them. It all adds to the sense of placelessness about the place; as though the researchers' wing was installed some time ago with the thought of being temporary, but then stayed.

Polly's office reeks of mothballs. Really reeks.

'Excuse the stink,' she says as we walk in, explaining that in decades past another means of preserving specimens was to use mothballs – and many of the display cases here are still full of them. She levers open a lid and runs a finger along the inside edge where a small slit has been cut and stuffed with tight white beads. 'The smell gets into my hair and clothes,' she says. 'I get funny looks on the bus sometimes.'

She heaves a pile of paperwork from an office chair and gestures for us to dump our bags. 'Ready? Let's start with the archives.'

At the other end of the corridor a small door leads through to a museum antechamber. It must be at least three storeys tall, a small square room with a winding staircase and bookshelves reaching almost to the ceiling. As we climb, the books are fewer, the shelves lined instead with old or misplaced equipment. There are test tubes, an old measuring cylinder, a desk lamp with the plug missing. And a thick layer of dust over everything. At the top is a small trapdoor; we clamber into the attic.

The attic is vast. It shines with deep mahogany. Great rafters lift upwards at the ceiling, and the low sun comes almost horizontally through the skylights. Rows of wooden cabinets run the length of the space, each one locked and labelled. 'Go ahead,' Polly says, handing us a key each. 'Take your time.'

Charlotte's off. She has a list of wasp and hornet species she's been trying to track down, and this is her chance. I move more slowly, and since I don't know any of the Latin names I begin opening doors at random.

Inside each cabinet are a dozen glass-topped drawers. I pull them out and find butterflies, beetles, ants – each one held in place by a pin pushed through the abdomen. They're arranged in rows, the pins an equal distance apart, with neatly handwritten labels underneath giving names and dates.

There must be thousands of insects up here; whole centuries of collecting and safekeeping. I imagine the team of researchers downstairs and all that work of classification happening. But – I realise – when I look at these specimens in their glass-topped cases I don't think of *order*; and despite the archive's promise of preservation, they don't seem entirely fixed. There's something less than precise about them; something more than category and type. They're different sizes and shapes. Some have lost wings or feet. There are bodies curled in on themselves; others that are stretched out almost flat. Each one seems a little skew-whiff. They're individuals, as well as species – and more fragile than I'd thought.

'Sometimes these pins are all we have to go on,' Polly says, peering over my shoulder as I pull open a tray of hawk moths. 'When we're dating a specimen, I mean. Handmade, and it's likely to be pre-1800. Copper, and it's an early machine-made pin. There aren't many of those left. Copper reacts with fats inside the insect, and the bodies tend to explode in time.'

Here is a name I recognise. *Apis mellifera*, the western honeybee. I pull open the drawer and crouch down until my eyes are level with the glass top. Each one has been pushed halfway up the pin-shaft – from this angle they could be hovering in mid-air.

It's strange to think that in a few weeks it'll be a live colony I'll be looking in on, crawling over each other and hatching or dying by the second. These creatures and those seem an impossible distance apart.

Glancing down the row of wooden cabinets, I wonder where boredom factors for the people who work here. Doesn't all that business of naming things and putting them into boxes get a bit tedious after a while? Doesn't Polly ever feel – trapped?

'It must take ages,' I say.

'Yeah – bloody painstaking sometimes.'

I ask how long she's been working here, and she closes her eyes a moment.

'Since I left university,' she says. 'God, nearly twenty years. Time flies, doesn't it? Blink, and suddenly there you are with a husband and a bloody mortgage and a kid all ready to start school.'

'But you enjoy it? The work, I mean?'

'Of course!' And, as if to prove it, she lifts a corner of her blouse an inch; lowers the buckle of her jeans. '*Shh*,' she grins, glancing over her shoulder as though there might be a museum guard about to appear and glare disapprovingly at us.

It's a tattoo. Just below and to one side of her belly button. An ant. The outline blueing now, the edges softening into the skin as they do after a while or after a belly has stretched and then contracted again.

'My first love,' she grins. 'Sad, isn't it? Sixteen years old and I was crazy about *ants*.'

She tucks her blouse back in and tells me how she used to track bugs and beetles through the garden as a child, mapping out their nests. Once she even bought an old fish tank in a charity shop and filled it up with soil, thinking she could create a whole ecosystem – an insect kingdom – and keep it in her bedroom.

'I spent days collecting bugs and beetles to put inside it. Ants too, of course. You should've seen the bites up my arms.'

'And did it work?'

'What do you think?! I fed them on kitchen scraps. It wasn't good. The scraps went mouldy, the lid didn't fit properly and most of them escaped. I didn't really know what I was doing.'

I used to go hunting for insects too. I had a jam jar with breathing holes punctured in the lid. I found a caterpillar once that made a chrysalis and turned into a butterfly, but it didn't do very well as a butterfly. Its wings were shrivelled up like crepe paper, and though I tipped it out onto the garden lawn it couldn't actually fly anywhere at all.

It crept over the grass, its antennae crooked, too sensitive and too flimsy for the world. I couldn't decide which was worse; to return it to the jar or leave it where it was, in the hope its wings might unfurl a bit. Then the wind picked up, and the butterfly went with it.

When we get back to Polly's office it is after six o'clock, and the desks are still humming with computers. She finds two microscopes and sets them on a table. Next she brings out a tray of specimens. 'These are spares,' she says. 'Duplicates, or species that aren't important to research. We keep them for visiting groups.' She puts her elbows on the desk, pushes the tray towards us. 'Take your pick.'

Charlotte finds one right away – 'A jewel wasp!' – but I'm still picking my way through the spread of inert creatures when hers is already under a microscope.

'How about this one?' Polly says, putting on her glasses and picking out a dark and curled-up creature. 'A honeybee. What you were looking for, right? This one's a female.'

She (the bee) seems tiny. But perhaps this is because of the huge beetles sitting to either side of her, who have messed around with my sense of scale. Polly switches on my microscope and the bulb beams a bright white. 'Put your eyes here,' she says, and I do, but my eyes must be too far apart because they don't fit the lenses properly. I try pushing the eye-pieces closer together, but then they pinch my nose. The microscope is heavy, and surprisingly bulky; I'm uncomfortable, and I don't see anything at first.

'Have a play around,' Polly says. 'You'll get the hang of it.'

Charlotte is well-versed in all this, and chats easily with Polly as she fiddles with the dial on her microscope.

'Have you travelled much?'

'I did a research trip to a rainforest in Bolivia,' Polly says. 'We were looking for wolf spiders; you see them best at night. Sweep a torch over the path, and you realise there are hundreds of green eyes looking at you.'

I seem to be looking at a shadow. Now and then I catch a rim of light in the lens but then it disappears, and everything feels back to front because I keep catching my own eyelashes in the glass.

'Scorpions, too,' Polly's saying. 'They glow a bright blue-green under ultraviolet light. With a UV torch you begin spotting them all over the path.'

I begin again, looking with just one eye and then both together. I catch a blur of something, and then it's gone.

'Actually, that was a very recent discovery,' Polly says. 'Until a few years ago, scientists couldn't understand why scorpions had evolved to light up under moonlight. I mean – they go hunting at night. And they're vulnerable to predators, too. Owls, rodents, things like that. So it seemed counterintuitive. Why become *more* visible, just at the point that you most need to work undercover?'

I seem to be struggling unnecessarily. I should slow down, I think. I try moving the bee, and not the lens, and catch a glimpse of that blur again. Keep my eyes fixed on that position; begin moving the dial very slowly towards me. I am turning it by increments so small that I am not sure I am really turning it at all.

'Three or four years ago they worked it out,' Polly's saying. 'Scorpions are extremely near-sighted. They struggle to make out anything at a distance. So they've learned a kind of trick, I suppose you could say. They use their bodies as light-collectors, taking in the dim light from the moon and stars and converting it into that bright blue-green that they can see. By sensing how brightly their own body is glowing, they can tell how much light there is in the surrounding environment. Like that they can find areas of shelter, even at night. They use their body as a gauge for what's around them.'

Suddenly the blur springs into focus, and my breath catches as I see her.

For a moment I think there's been a swap, she is so different close up. The plates on her back are black and tarnished and they glint like a pool of petrol. They're punctured with tiny holes. Her abdomen is furred, but the fur is matted and mangy-looking, like a dog that's been out in the rain. I look at her and think of metal and petrol pumps and the weather; not of bodies, or of bees. And then she dips back into dark.

I look up; adjust my seat. There's a message board on the wall still pinned with photos from the staff Christmas party, and a reminder about how to retrieve articles of lost property. Polly is picking through the tray of duplicates and discarded specimens. 'Some scientists even think scorpions have clusters of nerves spread throughout their body, which can register UV rays,' she says. 'If that's true, it would mean they're not just watching themselves glow; the body itself becomes a giant eye – *it* begins doing the seeing.'

I look down again, and see her.

On top of her head are three raised spots like black and shining stones, the kind you find on the bottom of a riverbed and take home for safekeeping, which dull as they dry out.

My mind is spinning out on associations. I try keeping it in, coming back to what I'm really seeing. My mind spins out again.

It's dark outside by the time we leave, and the moon is almost full. On the cycle home I half-close my eyes until

I can't make out shapes, just the blur of road and buildings and the feeling of air passing. I am trying to sense my own luminescence. It is not really the kind of thing that a rational and dispassionate observer would do, and I nearly crash into a lamppost.

'Steady on!' A man on a bike behind me slows as he approaches. 'You okay?' I nod and try to look responsible and roadworthy, and he smiles and rides away. There are fluorescent strips around his ankles and on the back of his coat, and they light up each time a car passes.

I cycle the rest of the way slowly and with both eyes open, thinking about the entomology department with the screened-off rooms and about Polly with the ant tattoo and all that knowledge neatly categorised inside her. I went because I wanted to know what it's like to bring a scientific gaze to a creature as curious as a honeybee, but I'm not sure a purely scientific gaze is what I found. Polly was passionate, and mixed up in her work. She was a scientist and also a mum and a wife, and once she was a child who collected a fish tank full of ants so they'd be right beside her when she woke. I'm becoming less and less convinced that it's possible to be a purely rational and detached observer of the natural world – or if that's all we'd want to go on, if it were.

\*

A week later it is still too cold to collect the bees. If a swarm has passed, it hasn't opted to stop here, and the hive sitting out by the far fence begins to seem to me like a half-finished gesture – a hand opened, unanswered, and left unretracted.

I remember Luke's approach of letting the weather decide, and try to keep my patience. Still, I can't help wondering if there's something more I should be doing to prepare myself; if there aren't rites of passage to becoming a beekeeper, things that have to happen before a hive – or indeed its keeper – can host a colony of bees.

Rained in, I begin reading about old rituals and traditions employed to attract wild colonies, or at least lend a bit of distraction from the wait. Across the ancient Mediterranean there was widespread belief in the practice of *bugonia*, the spontaneous generation of a swarm from an ox- or bull-carcass. The *Geoponica*, a Roman compendium of agricultural lore, lays out detailed instructions for this task:

The beekeeper should find a building ten cubits high and the same in breadth, with equal dimensions on all sides. There should be one doorway and four windows, one in each wall. Next he should *drive an ox into it, thirty months old, well-fleshed, rather fat*, then call a number of young men and have them beat it with bludgeons until they kill it.

Every aperture of the animal should then be stopped with cloth, including the eyes, before the door and every window of the building are closed and sealed with clay, *so that no air or wind or anything else can get in or ventilate it.* After three weeks all entrances should be opened, and light allowed to pass through until the room and all substances inside it are *sufficiently aerated*; then the door and all four windows should be closed and sealed again with clay.

After eleven days the room should be opened up, whereupon it will be *full of bees swarming on each other . . . and nothing left of the ox except the horns, the bones and the hair.*

I gaze around at my own room, imagining those windows and that door opening, the walls inside all thrust with light; wondering if bees really were found clustering, or if they were just flies feeding on a decaying carcass.

And then what am I looking at? I am looking at a bee.

Alive, and crawling very slowly across the carpet.

If you were to take out a knife and open her body up, you probably wouldn't recognise very much. A honeybee has an open circulatory system; the blood isn't confined to arteries and veins but instead fills the body cavity, surrounding her organs in the form of that yellowish liquid you'd see oozing if you poked.

But now look at her heart. A honeybee heart has five openings, each with a one-way valve; not something the bludgeoners or anyone else in ancient Rome could have known, in the absence of microscopes. Yet when I think of it now, when I picture a bee with her body opened, that account of the room with the four windows and door appears suddenly as an image of her heart. As though the room were embodying the bee, even bodying her forth; as though the wild, amorphous mass of a colony might be summoned when called by the bee's very physical form.

The real bee is still journeying her way across the carpet. I shepherd her onto a piece of paper, open the window and nudge her out. She dives, recovers herself, and disappears in the direction of the traffic.

The glass in the window is thin and the frame has seen better days; even when it's closed I can sometimes feel a draught in here. But there must also be a door or a window open elsewhere in the house because here are goose pimples on my arms, the tips of my hairs bristling, my heart quickened and my skin suddenly sensitive to the touch. *Bizarre*, I think, and open the window again, so as to give it a good slam closed.


And then just like that – before I can think any more about rites of passage or sealed-off rooms or what to do, how to be *really ready* for the bees – the weather makes a snap decision. Spring arrives, over the course of a weekend.

On a lunch break I find a quiet side street and sit down on a doorstep, fishing in my bag for the crumpled Christmas card with the pencilled honeybee on the front, my friends' names written on the back. I pull it out and smooth the crumples free. Beside their names is the address of a farm near Banbury, and a number to call come spring.

The gift was Becky and Jack's idea, and they'd heard about the farm through friends. It's run by a couple named Lucy and Viktor, who breed honeybees and sell them, and they sell honey too. Over the winter, at Becky's request, they've been keeping a colony aside for me; and when Lucy's picked up the phone and worked out who I am, she says they normally open for collections on a Saturday but in fact it's okay to come anytime, as long as this good weather holds.

So I call Luke to tell him, and he says how about this weekend; he can come down from London and we can drive out and collect them together. We make some arrangements, and sign off. Suddenly it is all really about to happen.

On Sunday I meet Luke at the train station and we pick up a rental car.

'I've been dreaming about hornets all week,' I tell him,

fiddling around with the satnav. 'And bees. I keep dreaming they're in the house, or in a box, and I'm looking around for my beekeeping suit, but I can never find it.'

Out of the city we pass through villages of neat thatched cottages, their windows squat and low and portioned out into dozens of diamond panes. A 4 × 4 in every driveway, and not a creature in sight.

'You want settled?' Luke says, because I've just been telling him about my seemingly ill-fated attempt at settling down in Oxford. 'This is settled.' He gestures to a hedge that's been topiaried into a peacock. 'I think I'd feel trapped.'

Beyond the villages are country lanes surrounded by acres of fields. It's all going well until the satnav fails, then we're on our own in an ocean of rolling green. Luke pulls up in a lay-by as I reach into the back for the hand-drawn maps and scribbled instructions we brought with us as backup.

'We're supposed to follow the signs,' I tell him. 'Lucy said follow the signs.' We follow the road until we've definitely gone too far, and then we follow it back again. Everywhere is a haze of field and sky. At one point we think we see a hive ahead, but it's only a fencepost.

The honey farm is a small island of concrete and corrugated iron in the undulating green, and we reach it only after Viktor has driven out to come and find us. There's a deeply grained chestnut tree beside the gate, its tips just beginning to find their leaves, and a row of weather-warped beehives at its base.

'We didn't see any signs,' I say to Viktor as we clamber out of the car. 'Lucy said follow the signs.'

He shrugs. 'There are only signs on a Saturday.'

The yard is scattered with pieces of farm equipment in various states of disrepair. There's a barn and a row of crumbling outbuildings with their windows painted over so that I can't tell if they're in use.

'Where I keep my honey,' Viktor says, seeing me eyeing a large shipping container. He points to a sturdy padlock on the front: 'For the thiefs.' He has a Ukrainian accent so thick that I wonder if he's putting it on. He's wearing a full-body beekeeping suit with the gauze hood flipped back over his reddened face, and the front unzipped to the waist. It's thickly spattered with wax and pollen, which gives the impression of some kind of slaughter having taken place, or of an oversized play-suit. 'So, who's the beekeeper?' he asks, looking us up and down.

I point to Luke, and Luke points to me. 'She is.'

Viktor turns and disappears into a shed.

'You want tea?'

The inside of the shed is piled high with pieces of hive, and it stinks of wax and wood. There's a desk at one end where a man in glasses and a boiler suit raises his hand to us, and a set of shelves with a kettle and a microwave at the other.

'Honey?' Viktor pours tea into brown-rimmed mugs and drops a large spoonful in each.

'How long have you been keeping bees?' I ask, looking up at a poster of a shiny motorbike leaping over a mountain ridge, and wondering where Lucy is.

'Since I was birthed.'

'His parents were bees, weren't they, Viktor?' shouts the guy in the corner.

'Shh,' Viktor says, and smiles to himself. 'Don't tell my secrets.' And then he gestures to me.

'Come – we get your bees.'

Behind the shipping container the ground is rough and bare and there's something in the air that catches in my throat. Perhaps it's pollen drifting over from the fields.

There's a wooden box between us, closed. We're both cocooned in suits now, with the hoods up and the gauze masks down so that I can't see Viktor's face any more. All around us more boxes are arranged in neat rows, stained rough reds and greens and looking a bit like improvised towers in a miniature and makeshift city. These are *nucs*. A *nucleus* is a small colony created from a larger one, containing a queen and a body of workers along with a series of wooden frames filled with eggs and larvae and honey stores. Before I take this one away with me, Viktor first wants to open it up to show that it's healthy inside.

'You're not wearing gloves. Won't you get stung?' His hands are red and swollen like his face was when I could see it.

'Me and bees; we same blood.'

As he takes a hive tool from his pocket and prises the lid open there's a sound like a thousand nerves tightening and stirring. He moves fast, without stopping. Takes a few frames out and empties them by jerking, so that the bees change from clinging solid to thick dark liquid pouring back down into the hive.

'Queen.' He points. 'See?' I spot her through the throng, bigger than the others, and shrinking against the comb. The comb is stuffed with unhatched eggs and I see pollen too. *My bees*, I tell myself, testing how it sounds. *This is the colony I'll keep*, I try. The one I'll get to know.

But we're disturbing them. Dark points of agitation fly up, away from the hive or straight for us, and my gauze mask thuds as one hits and holds, buzzing. I can't see them separately any more; can only feel the size of the disturbance spreading, a hot low pulsing that swells until it surrounds us, and then we are inside it, and the air is alive with them. I'd forgotten this. Had forgotten the for-eignness, the formlessness, the furious buzzing that comes sometimes as you lift the lid, and their rising that is more like heat or sound than movement.

Viktor's finished; he replaces the lid and shuts the entrance closed. Except for a slim metal ventilation grille at the top, the box is sealed. The hum quietens, cools, stills. There is a lot of light around and something is ringing, like when you've gone underwater and come back up and everything seems louder than before.

'Here,' Viktor says, lifting the box and gesturing me closer, 'take it.'

Two crows heave from a tree and a scatter of deadwood falls. Somewhere a generator whirrs. A gate swings open on crusted hinges and clangs, repeatedly, against a metal fence post.

The box is heavy. Heavier than I thought. My arms prickle at the thought of all those bodies freshly upturned and moving around inside. Checking themselves, checking each other, sensing walls. I feel a bit light-headed. I turn, blinking, to see Luke waiting by the car with the boot open. He's hopping lightly from one foot to the other, impatient now to go.

We put the box in and pad it with coats to keep it from bumping around, and we get in the front. 'Good,' says Viktor, nodding, and gives a wave as we inch down the rutted track. Then he turns and disappears back into his shed.

A few weeks later Lucy calls to ask how the bees are, and we end up chatting about the farm and about how Viktor came to be a beekeeper in the first place. No one in her family kept bees, she tells me, but his whole family were beekeepers; he grew up in the Ukraine during Soviet times, when you were supposed to have just one hive per household, but Viktor's dad had over twenty hidden in a nearby forest. Because of this the family were quite well off, and Viktor was able to go away to university in Kiev, living high up in a block of flats, which was okay at first, but after a while he got homesick and he missed the bees, so he bought a hive of his own and kept it on his apartment balcony. This helped a bit but not enough, and eventually he gave up his studies and went back to the family home. He's been a beekeeper ever since. When winter comes each year he is glad to have a rest, but by January and February he gets sad and listless and struggles to put his mind to anything, until the spring comes when he can be out with the bees again.

As we drive back to Oxford with a box of bees in the boot I try to find words for the feeling. 'It was like – a warmth. Like there was a warmth coming from them. Have you ever felt that before?' I ask, and I keep twisting around in my seat to check the back window because I'm half-expecting to see a colony of bees flying out.

Luke isn't into mysticism but he gives room to experience, and he likes to wonder about things. 'Well, it would've been warm in the hive – the bees have to keep it at around thirty-five degrees for the brood to develop, and it's cool today; you might have felt heat escaping?' Which makes sense, but doesn't fit with the feeling. It wasn't coming from the hive itself, it was coming from the space around me. Perhaps *warmth* isn't right; I can't match my words to it.

When we stop at a service station I open the boot and put my nose up to the metal grille where upside-down-bee feet are clicking and tapping. They're *beautiful*, I want to say. I am *amazed* by them. I say it to Luke when we're driving again, but my eyes are too wide, it sounds overblown and silly. That multitude of tiny legs, each one no more than a hair's breadth. I can't wait to get them home.

Back in Oxford we put on our beekeeping suits, then take the box out and open it, and the bees lift up as we slip the wooden frames out and place them inside the waiting hive. The opened cavity seems impossibly vast after the tight-packed box, and the frames don't fit; they're too square and wide to hang widthways down the tapered sides, so instead we rest them like fallen dominoes along the base. I catch sight of the queen, then, before replacing the lid.

That week it is suddenly cold again. So cold that people stop in corridors and on buses to mention it. 'So strange,' they say. 'What's happened to the spring?'

I walk around the office with a lightness in my stomach, pausing at windows to peer out at the roiling clouds, and then it rains, a lot, and I worry the bees won't get out to gather food – and what about that steady inside temperature they have to keep?

I begin checking online weather reports every few hours, and they warn of freezing temperatures. So there I am, running out at night under a bitten moon, heaving blankets and string from inside the house and wrapping them around the barrel of the hive. I don't open it up to look inside because I don't want to take any more heat from them.

Like this the hive is just as before, except now with those blankets around it, and the frames inside like an organ grafted. I watch it for a week. Not believing the bees will make it through, not knowing how they could – wondering at the shock and the cold and the violent displacement of arriving here, coming to this place.

4

# Orientation

# April

After those first few days of rain and cold the clouds begin to thin and then the sun bleeds through, and a flight path opens in the garden. Through the windows we catch glimpses of bees hovering around the hive; shuttling over the fence. Not many. But a few.

Now they've arrived in the garden I wonder how on earth to begin *keeping* them here. There's a terrible kind of fluidity about them with that open doorway and those passages back and forth. I realise I'm spending every second moment worrying they've died in there; every third worrying they'll fly away. *Enough*, I tell myself. Go out and look.

There's a tree stump a few metres from the hive entrance and this is where I sit as the air gathers a warmth to it, waiting for a sign or some indication that they might have accepted this tin-topped, too-bright thing. And they do get busier. I watch the way they sweep down and pause, holding themselves over the threshold before disappearing inside. Later I move closer and glimpse pollen like globs of earwax clodding their thighs. The pollen is bright yellow or sometimes orange, and they'll be collecting it to feed the young larvae hatching inside the frames. They might be drawing nectar, too, for making honey; but that's a thing that you can't see being carried, no matter how close you get.

If there's pollen going in, there must be *something* happening inside, I tell myself, but Luke said to leave them for a week so for the moment I'm only guessing. If all goes to plan they should start moving off Viktor's nuc frames to find the bars at the top with the neat wax strips, and building their own comb freeform. Once laid, eggs take three weeks to develop into female workers (a little more for drones). Workers live for as little as six weeks in summer and are crucial to the functioning of the hive; the colony will become weak and will ultimately die off unless the supply is constantly replenished. So I should wait a while before removing the frames, Luke said, his thinking being that by the time the eggs in them have hatched there should be enough new brood in the fresh comb for the colony to sustain itself.

There's a rustle behind me, and Hannah's face appears on the other side of the hedge. It's in leaf now, but still sparse enough to see through if you push.

'They're in?' she says. 'Wow.' She pulls at a branch so as to see better. 'How many are there?'

'I don't know. A few thousand?' I hope, eyeing the hive.

'How are they doing?'

'I'm not sure. We're waiting to see if they'll take to the hive – start treating it as home.'

I look around at the garden, which seems itself to be toying with the idea of domesticity. Becky's cleared the container beds and laid them with old patches of carpet to stop the weeds; she's mowed the lawn, too, with a rickety old lawnmower she found in the shed. The grass is roughly shorn and patchy – it looks as though it's had a rather difficult encounter with a hairdresser.

Three more bees lift up from the hive and disappear over the shed roof.

'Where d'you think they're going?' Hannah asks.

'They're probably scouting the area – looking for what's in bloom. They fly up to three miles when they're out foraging.'

'I wonder what they'll find.'

I try imagining the hive from above. With that tin roof the sunlight bounces so that it could be water or a window facing up. It is one rectangle sitting within the larger rectangle of our garden, which itself appears as one in a long line of other rectangles following the terraced houses on our road, and other lines of other rectangles, spreading out from the city centre and into each other in a great web of neatly lined and edged and compartmentalised space.

'Exciting,' Hannah says.

'Yeah.' Except for that pool of nervous anticipation collecting in the bottom of my stomach, and the buzzing it is making in my head. I am longing to lift that lid.


'A suspension of disbelief,' Ellie says on the phone that night, when I call up in a bother to tell her I'm convinced they'll either leave or die. 'You have to stop stamping it down so as to make some space for it to happen.'

My feelings are flying about all over the place. What's got me in a tizz is not the fear of failure or even added responsibility, but the sudden rush of *care*. I've put a lot of thought into how this will work, what makes a good beekeeper – I hadn't expected to *want* the bees; I hadn't expected to feel so implicated.

'Try not to get so caught up in things,' Ellie tells me gently, after we've said goodbye and before she puts the phone down. I'm not getting caught up in things, I want to say. Or I wouldn't be if they decided to stay. With the phone still crooked between neck and ear I open the curtains, peer out at the moonlit garden. A peculiar quality of moonlight: it stills things. It suspends them.

All the movement and change of the last few years – the string of houses and jobs, the roving friendships – has put me in the habit of expecting things to disappear quickly once found, and in that state of almost permanent temporariness I've caught myself wondering a few times if it's very reasonable to hope to keep anything at all. No wonder the arrival of a colony of bees is triggering my anxieties. Here I am pondering impermanence, having just tasked myself with the responsibility of *keeping* something – with sustaining it. A colony is not a book or an archivable object and you can't hold it in a glass cabinet or on a shelf. It is live and shifting and if this one doesn't take to our little rectangular space it'll be out of here faster than you can say *swarm*.

That glimpse of the bee under the museum microscope has been playing on my mind like an image caught and flickering. I'm falling asleep at night and suddenly there she is, springing into focus. Or there she isn't. I want to find her again.

A worker bee spends most of her life in the dark inside the hive, where there is little need for sight. She has five eyes. There are two compound eyes like giant cheeks on either side of her face, and three *ocelli* – light sensors – on top of her head (these are the spots like shining stones that I saw through the museum microscope).

Charged with almost total responsibility for maintaining the healthy functioning of the colony, her life passes through a series of stages, and in each she has a specific role to perform within the hive. She is first a cleaner and then a nurse bee, raising young larvae inside the comb (a single larva will be checked by nurse bees over a thousand times a day). Later she becomes one of the queen's attendants, and also takes a turn in keeping the hive at a steady temperature, fanning her wings to keep it cool and ventilated. The stages are pre-programmed to the extent that her bodily capacities change as she matures: at seven days her glands begin secreting wax and she becomes a comb-builder (or she becomes a comb-builder and she finds the wax secreting). New capacities emerge, forcing her on, out. She leaves the hive, first as a guard bee and then as a forager, only towards the end of her life.

So when I sit out by the hive, these are grown bees I'm seeing, already familiar with the internal rhythms of the colony. I'm just back from work, and slowly coming to – my eyes still strained by screen-glare. A bee turns and lands on a dandelion beside my right toe; I bend down for a closer look.

I can see her compound eyes but I can't see her light

sensors, which are lost in a fuzz of hair over her face. The fuzz of hair is how she senses which direction the wind is blowing, and her own flight speed. *Ocelli* don't register images, only changes in light intensity, and thus where the sun is in the sky. Knowing the sun's position tells her which way 'up' is, which brings stability in flight, and helps her to orient herself.

She thrums her wings, and the dandelion bobs a little as she lifts. I twist, fidget, scratch my head. I wish I were able to learn 'up' and a bit of steadiness just by looking at what's around me.

I peer at the hive entrance and think of Aristotle and the other classical greats; all the things they guessed and gleaned about the life of the hive in the absence of windows or lids. I've been sitting and listening and waiting and watching, but don't seem to have gleaned anything so far.

She might have those sensors like an inbuilt compass directing her to use the light as a point for anchoring, but a honeybee's relation to the sun itself is active and learned. In their book *The Honey Bee*, James and Carol Gould describe an experiment by the scientist Martin Lindauer, who raised a colony inside a closed room and then transferred the hive outside. The bees quickly became lost. A fixed light bulb proved an inadequate substitute for the sun, which *moves* – and without having learned to orient themselves in relation to a light source that rises in one place, dips in another, and sometimes disappears altogether, they were unable to navigate a landscape or re-find the hive, having departed from it.

'I feel like a doughnut,' my friend Kath says, mashing a poppadom. It's Friday night and we're sitting in the window of a greasy curry house behind the train station. 'I'm working four jobs and they're great but sometimes I feel like all I'm doing is moving between them, running round and round, I am so *busy*, and in the middle is just a hole. That's when I feel like a doughnut.'

She sprinkles a plate of curry and rice with the poppadom crumbs and begins mixing them in with her fingers.

I'm about to say that I don't know if the doughnut is a cushioning or the stuff of life itself, but then the waitress arrives with a water jug and plastic cups and a cloth that she fusses the table with. She's wearing a purple baseball cap and a matching polo shirt, and when she grins a gold tooth flashes. She's come from Poland, she says. She wants to practise her English, but working here she ends up speaking Hindi all the time. She rolls her eyes, flicks the cloth, and disappears back into the kitchen.

The poppadom crumbs have gone soggy from all the

mixing. 'There's a lot of loneliness around, I know that much,' Kath says, forming a small ball of coagulated goo between finger and thumb and popping it in her mouth. 'A lot of instability. Sometimes I'm not even sure if it's about trying to get anywhere any more; I'm not sure where I'm coming *from*. I run around in circles, chase after things. I start thinking I need to socialise more. I socialise more, and it just makes me tired.' She sighs, and laughs. 'I need a holiday. I can't afford a holiday. I fucking want one.'

There must be other ways of doing things, I think to myself. Some patterns beyond the ones we're used to seeing. Outside the window a busy junction streams and plugs with the last of the rush-hour traffic. The sun has dipped down behind the office blocks opposite, and the blue is draining from the sky. I'm remembering that original meaning of *home* as a place of sense-making, of world-building; a place from which journeys are made.

The waitress reappears at the table next to us, balancing plates from hand to wrist to the inside of her elbow. 'Hot!' she says as the plates go down. 'Hot, hot!' The plates are steaming.

'Anyway,' Kath's saying. 'How are you?' And by way of a reply I do a bit of complaining too, about work and stress and feeling pushed. 'You're too sensitive,' she says, and wipes her nose, 'that's your trouble.' Which doesn't help because I've been trying to adjust by toughening up; but perhaps I am not as self-sealed as I'd thought.

Kath reaches for the water jug. 'Anyway,' she says. 'How are the bees? They've arrived?'

'Yeah, they're okay. I think. I mean – I haven't opened the hive yet.'

'Funny, me going on about being so busy,' she says. 'What about the bees? Now *there's* an example of a tough work ethic.'

I drain the water from my plastic cup and sit up in my chair. Now *here* is a thing I can speak about. 'Well,' I say, 'you'd think. But actually bees spend about eight hours a day being pretty aimless. Just wandering around in the hive, not doing anything very much.'

'Ha! And I had them down as a model of productivity.'

'Well, they're that too. But the wandering around is all part of it. They know to preserve energy for when they need it.'

Behind her head, beyond the window, the car headlights blink. On this side of the road there's a row of shops, and you can watch how the light in people's faces changes as they walk. A harsh white outside the brightly lit kebab shop next door, then a soft orange glow at the bar a few doors down, and a flickering green or pink and blue outside the newsagents on the corner where *Lottery* and *Phone top-up* signs flash through the glass.

'So, when will you do it?'

'Do what?'

'Open the hive.'

'Oh – at the weekend. Tomorrow. If the weather holds.'

The next day is Saturday and the weather does hold. It is one week since the bees arrived, and time to make the first hive inspection. Most beekeepers will inspect their hives once a week in summer. You lift the lid and take the bars out one by one, keeping an eye out for signs of disease, checking for brood and pollen stores, and that the queen is alive and well. This first inspection of my hive is a bit simpler than all that: I just want to check that the bees have moved off Viktor's nuc frames and begun building their own comb from the top bars.

I pull on my suit. Big boots, thick gloves, a gauze mask pulled down over my face. There's a hive tool in my pocket like a miniature crowbar painted yellow, ready to split things apart. I hear birds and the traffic, a guy spitting and shouting in the street, the sound of my own head brushing the inside of my hood. No sudden move-ments, I know that – but it's not easy to walk inside this suit. I shuffle closer, not up to the entrance this time but at the side, in the middle, where I stand with both hands on the roof.

'I'm going to lift the lid,' I tell them. I say it quietly, so as not to surprise them. 'I hope you'll stay,' whispering this, so that even if there were someone listening through the hedge or over the fence they wouldn't hear me.

I squat a bit, to get some leverage. Those mismatched nuc frames are in the middle, at the base, about level with my navel. I am standing with my arms outstretched; I clench my hands and lift.

No sound. Here is the row of top bars, in the place where the roof was. I adjust the gauze, away from my face. Begin at one end, teasing a bar up, out, feeling for a weight, something stuck or building underneath. It's empty. Clean as it was when it first arrived here, except for the neat wax strip that I melted on. Next one: nothing. I move through the bars, one by one, lifting, tensing for the feeling of something added. I reach the other end of the hive; the bars haven't been touched.

Where are they?

Step back.

I check the entrance; see a few bees drifting and pass-ing. Bolder now, with worry, I remove four bars at once, making a hole big enough to peer inside. And then I see them. Crammed into the nuc frames like a tight and con-centrated knot sitting fixedly at the bottom. Except for a few foragers shuttling back and forth from the entrance there is no sign of any movement *out*; nothing to suggest they've begun exploring the waiting cavity, let alone accepted it as a home. In fact, as I take the hive tool and run it along the walls, I find propolis gumming the nuc frames; the bees have welded them further *in*. I close the gap, replace the roof. Retreat to my own doorway.

'Nothing?' Luke says, when I call him.

'Nothing.' I watch the hive with half an eye as a bee lifts up and out of it.

'Nothing at all?'

'Nothing at all.'

'Give it another week,' Luke says. 'If they haven't started building comb by then, you might have to take drastic action.'


I do as he says. It is warm enough in the mornings now to sit on the tree stump before work, and I begin taking a cup of tea out in my pyjamas to watch the first flights tipping and leaving. The day's journeys begin when the sun reaches the hive mouth, and they end when it lowers behind the fence.

I don't lift the lid; and I try to leave my disbelief suspended. When the mug is empty or the tea is cold I go back inside, get dressed, get to work, return to my own frames and schedules. I try not to get caught up. I run through to-do lists, answer emails, attend meetings. But

I am thinking about them. The thought is like a breath taken in and held somewhere behind and underneath my ribs. My insides are light like the inside of the hive is light; I'm waiting and willing them to stay.

When I get home I go outside again, try to keep sight of one bee long enough to know her. I can't. If I could, I might ask her why it is that she won't move off those frames.

Then one day an email arrives from Jack with a picture attachment. I'm sitting at my desk at work, and really this is not the time to open it. There's a lot to do, and I don't need any more distractions. *Closer than a microscope!* is the subject heading.

And it is. Zoomed up, in special, high-definition focus that makes everything appear smooth, I find her face. I glance up, around the room, to see if anyone's noticed that I have stopped looking at my to-do list, and begun looking instead at her face.

Here are two raised mounds, in the place where our eyes would be. Softly furred, and hard like the welts on the heads of young cattle at the point where their horns will be. In place of cheeks are her compound eyes, round and so huge that her vision extends to almost a complete sphere.

Her compound eye is a convex surface of thousands of individual lenses, each one independent of the next and angled slightly differently to cover a unique part of the visual field. There is no lens at the back, as with a human eye, which works to gather an image from the information coming in and so form our perception of a continuous visual field. What she sees is a patchwork pieced together from those thousands of angled views. Highly pixelated, we'd call it. A low-resolution image. If a bee were to take an eye test, by our standards she'd be classified as almost blind.

I do a search on Google Earth for our postcode, wanting to get a sense of what the bees might be seeing as they fly up from our terraced plot. The web page is slow to load and it's difficult at first to make out an image on the screen. But this, right now – as the picture is still assembling itself – must be something close to what she sees. Grainy blocks fuzz and form across the screen.

Objects on the ground are visible only if they're big, or she's close. Tree, house, chimney pot. The landscape becomes distinguishable, and so also navigable, by a series of features: *things* that rise up out of the blur and prove large and immovable enough to mark a flight path. Like this, it seems impossible that she can navigate at all – but then she can also see things that we can't. Like polarised light, which forms a particular patterning across the sky in relation to the sun, such that, by watching for these patterns, she can tell where the sun is even when it's lost behind clouds. Or like ultraviolet light. I find pictures online of flowers photographed under a UV bulb and see marks on them in places I hadn't noticed before; colours that contrast where I'd thought them uniform. There are lines like arrows directing bees to their centre.

Later, as I step out of the office onto the street outside, I look up; and for a moment I see patterns forming themselves. Look down; and imagine markers, and signs. The route home takes me down pavements and across roads. There is a road sign every few metres; a traffic light at every turn. It is all a long way from the experience of a honeybee, who adapts to her environment rather than writing over it, and whose systems for navigation are stowed inside her.

The next day I'm outside the Turkish shop on my road when a man with a streak of grey in his hair and a swollen lip reaches across me for a courgette. I've been sorting through tomatoes, searching for a bunch with the vine still holding them. I like it when the vine still holds them. Sometimes I bake them in a bunch like this, and then the vine goes dry and crispy like a piece of twisted seaweed. I'm not sure if you're supposed to eat the twisted seaweed, but anyway I do.

'Bee sting,' he says, seeing me eyeing the lip.

'Oh,' I say, and wince. Swollen lips always feel bigger than they look, and this swollen lip looks big.

'Do you have a hive?'

'Five,' he says, drawing himself up. 'Out of the city.'

'You weren't wearing a suit?'

'I was wearing a suit. But I wanted a closer look. Held the frame right up to my face, and one got me through the mask.' I push the skin of a tomato with the tip of one finger and watch tiny wrinkles form over the flesh. I can't help envying him his frames, all lined up and laid out ready. None of the uncomfortable mismatching happening inside my own hive, or that empty vastness of the waiting cavity.

'I have a hive too,' I say, pointing up the road.

'City's a good place for them.' He nods approvingly. 'Plenty of biodiversity, see. Not like where I live. Oilseed rape – that's all I get. It comes, and then it goes, and then there's nothing. Huge fields; no hedgerows. *Pffllp*.' His mouth makes a sound like a balloon deflating. 'I wouldn't like to be a bee now.'

Underneath the tomatoes are trays of oranges and lemons and one of scotch-bonnet chilli peppers, making bright blocks of colour against the grey of the pavement.

He's right about the rural landscapes. Over the last century there's been a massive drive in the UK towards intensive agriculture; farming practices that inadvertently supported pollinating insects – the hedgerows, fallow fields, and even the alfalfa crops which provided fuel for the horses that pulled the ploughs, while also providing forage for bees – have been replaced by systems of large-scale mono-culture and agrochemical use that are harmful to them.

'*Habitat fragmentation*,' the lip is saying, picking a cucum-ber up and waggling it at me. 'I went to a talk about it. Some guy from the university, don't remember his name. Pollinator habitats are getting shrunk into pockets. Like *islands*,' he says, and puts the cucumber back. 'And the thing is,' he adds, looking down at the tray of them, 'the thing *is*, the islands don't join up; the bees can't move between them.'

'Hey man, are you gonna buy those vegetables or just stand around playing with them?' The shop assistant leans over the counter, then baulks at the sight of the lip. 'Oooh.' He winces, and makes a face. '*Naaasty.*'

As we pick up our baskets and take them over to the till I see this place from above, again. Roofs and roads. Those fenced rectangular gardens with the different-coloured flowers inside them, and the pockets of wild space between: a corner at the top of the recreation ground near our house that's begun filling with wildflowers, and the crack between two walls where last week I spotted a sprig of buddleia. A patch of land beside the golf course where

Becky found wild raspberries growing, and a verge spilling weeds beside the train tracks. I see all this, the shoots of weed and wildflower mixed in among neatly laid areas of cultivation, and then I see the flight paths looping and linking them.

'You okay?' asks the lip, seeing me wobble slightly.

'Yeah.' I blink and shake my head. 'Bit dizzy.'

The shop assistant is half-hidden behind packets of chewing gum and KitKats. He's weighing out vegetables, and in no kind of hurry about it. The woman in front is buying yams. Her kid hides and holds between the folds of her skirt, watching.

'So you're pretty new to beekeeping, are you?' says the lip. 'How's it going so far?'

'I – they're still adjusting, I think.'

'Mm.' The lip stretches – I think this means he's smiling. 'I imagine you're still adjusting, too. Takes a bit of getting used to, walking out your back door and finding a bloody beehive sitting there.'

I grin. 'Yeah, maybe.' Perhaps keeping a wildness close doesn't always come very naturally at first.


The inside of the hive smells tart and tough like a hamster's cage, and there's another scent too, underneath the first, something sweet and soft and seething.

Saturday has come around again, and I'm making my second inspection. This time as I lift the bars up one by one something sticks.

I peer inside. Here is a lip of fresh comb running down from the top, but it's odd and misshapen – it twists awkwardly in the middle and then fuses to a nuc frame at the base. My breath catches and my chest chills; I've never seen anything like it. I close the hive up quickly, come away again.

'Take them out,' Luke says decisively when I phone him.

'The nuc frames? All of them?'

'All of them. You're going to have to leave the bees to go it alone. Either they'll take to the hive or they won't, but you're going to end up in a mess if you carry on like this.'

A colony of bees will always seek to build along a single orientation. Since the top bars are rested widthways across my hive, when the bees began building they met with confusion halfway down as they reached the frames sitting lengthwise along the base. A mess indeed.

'Just make sure you get the queen in. If she's there, the rest will follow.'

I pull my suit back on, zip up. And as I lift the lid this time my knees are light with a feeling like thieving. I prise the frames apart and there's buzzing as I lift one, take a breath and shake it like I saw Viktor do in one clean movement, except that my movement is not clean and the bees don't pour like a liquid, they just buzz and fly up and some of them fall back into the hive, but I don't know if the queen is among them. This is more like ripping than keeping, I think, as I pull the next frame out and shake it. Everywhere I look now, I am looking through a fuzz of them. I don't care, I think. Let them leave; I don't want any more of this. I dump the last frame by the foot of the hive and am about to close it up, get away from here, when I remember one last thing Luke said to do. I take a kitchen knife from my pocket, find the bar with the freshly built comb and slice it at the point where the twist begins; pull the warped part free.

'It's done.' I'm breathless and shaky when I phone him again, and my voice has a faraway sound to it.

'Good. And the queen's in?'

'I – think so.' There are still clouds of bees fretting the entrance.

'Good. Okay. And now you wait.'

Later and without gloves on I take the piece of torn-off comb and hold it in my palm. It is so light that I can't feel the weight of it. If I pinched one cell between two fingers I could crush it. I don't, but it unnerves me. I put it on the shelf in my bedroom beside the bottle of hornet whisky and there it stays as I move around, distracting me with its thinness and unfilledness and with the suddenness of its form.

That week I avoid the hive. I am busy. Out a lot, and when I'm not I steer clear of the windows. If I do catch a glimpse of the hive sometimes I see bees circling the entrance, but not many. Not enough.

Still, I can't avoid thinking about it. That big boat-like space with all the frames removed – will there be a reorientation happening inside? Will the bees move up, around, begin building again from the top? Or will they give up, go off, search out somewhere else?

We're hanging washing on the line one afternoon when I can't help but notice the lack of them. The sun is shining, and they should be out – but I see no sign of them.

'Have you seen the bees today?'

'No,' Becky's voice comes through a bedsheet. 'Not at all. Do you think something's happened?'

Well, I think. They've left. That's what's happened.

But a little while later we see a few.

'Funny, isn't it?' Becky says, standing outside the back door as I cook dinner in the kitchen. 'You think they've disappeared, and then there they are again.'

But I'll wait out the week before opening the hive

up, since at least not doing anything is a thing that I can do.

I am getting *way too caught up*. This much worry is out of proportion with a colony of bees that has just arrived in the garden; I need to get a grip. I retreat inside, defer to the dictionary. At least the dictionary is around to help make sense of things.

In the online Oxford English Dictionary the entry for the verb *keep* is long, and its etymology is curious. While over time *keep* came to imply (with increasing intensity) an effort to *retain*, its earliest meaning is likely to have been something closer to *lay hold, with the hands, and hence with attention; to keep an eye on, to watch*.

I shift my laptop over to the opposite knee; read through a second time. And then I take a breath. To think that keeping may originally have been associated not with locking down but with taking care, and that care was about eyes and hands and the very particular kind of attention that flowed from them. Well, that shifts my focus.

Since my friend Ellie works at the OED and knows a lot about words and their meanings, I write her an email to check I've understood it right, but she replies to say that I should take the definition I've read with a heavy pinch of salt. The whole dictionary is in the process of revision and *keep* hasn't yet been updated, which means it was last worked on more than a hundred years ago – in 1901, to be precise. *So basically the entry you are looking at is a piece of nineteenth-century scholarship which should be treated like anything of that age*, she writes. *Venerable but fallible!* But *keep* does seem to have a very wide spectrum of meanings, she adds, and goes on to make a list of all the things it makes her think of.

Here are the words on her list: *observation, care, nurture, interception, seizure, absorption, maintenance, preservation, retention, detention, restraint.*

I wonder what would happen if everyone in the whole world were to make a list like this. Sometimes our lists would overlap but probably there would also be a lot of differences, and each one would tell a story about that person's own experience of keeping and holding and being held. I write back to Ellie to ask when will *keep* be updated and she says she doesn't know, that hasn't been decided yet.

Perhaps it's no surprise that I've been feeling uncertain about how to *keep* if even the dictionary isn't sure of a right meaning for it. But then isn't our whole mode of keeping – our capacity *to* keep – undergoing a major crisis of confidence at the moment? Climates are changing; colonies are dying. Our ability to sustain environments of all kinds is being called into question, and we're having to face our own dependency on those ecosystems we've seemed intent on destroying in our bid to control and manage nature.

Quite apart from the dictionary update, taking pause to rethink and revise what we mean by *keeping* and how we go about it may be badly needed.

I can't open the hive the following Saturday because a group of friends are visiting. When they arrive they say how are the bees and I say I don't know, I took everything out from the middle of them, there's nothing holding them there now.

We pull out the table for lunch in the back room and gather enough chairs to seat everyone. While I'm in the kitchen fetching plates and forks, they pop out to the garden and come back telling me the bees are busy, it's looking good, there's definitely something happening.

The table fills with dishes. It is too small to hold everything, and we are too many to fit around it. Our elbows bump, and so do our plates and forks. We speak through mouthfuls, over each other, across the table. It is the kind of bubbly and excited chatter that comes after you've gone a while without seeing each other. For the most part I bubble along as much as anyone, but a few times my attention drifts. I catch myself glancing out at the hive, scanning the air for movement or some sign of life.

'They're going to be *fine*,' Dulcie says, seeing me looking. She has a dark bowl-cut and pink cheeks and when she feels strongly about something she gets so full of energy and conviction about it that usually, whatever it is, I end up believing her. 'The best advice I got before Corinne was born was never to rush her,' she says, bringing her face close and stretching her eyes wide so that I have to stop looking out at the hive, and see her face. 'Let her settle into the world at her own speed. *Relax*,' she tells me. 'Give them time.' It is easy to believe what someone is saying when you know that they are able to believe themselves.

Later I go out and see for myself. When you sit right beside the hive the bees sweep close to your cheek, but if you're quiet you aren't a bother to them, and this is how Laurence is sitting, with his legs crossed by the concrete slab, while everyone else is inside getting pudding.

'Is it okay?' He looks up as I join him. Laurence has blond hair and broad shoulders and he's sitting right by the entrance in shorts and a T-shirt, as the bees with their stings and shining-pebble eyes fly a few inches from his face.

'To sit here? Course.'

He has no more understanding of bees than the next person, and he's never been this close to a hive before, but he doesn't look the least bit uncomfortable about it.

'Amazing, aren't they?' He nods at the hive.

I follow his eyes for the amazing.

That last nuc frame is still lying where I abandoned it at the foot of the hive, and I notice that it's been sucked dry of honey. That's a good sign, that the bees have salvaged what was left and carried it back into the hive. I pick it up and put it away in the shed where the rest of them are stacked, then I come and crouch down beside him. We talk a little, but quietly, not wanting to disturb the bees or distract them.

And then he whispers, '*Look!*'

He's pointing to the foot of the hive, to the place where the frame was. One bee and then another is tipping down and hovering there, hesitating almost, before flying away again.

'They're coming for the frame,' he breathes.

And then I see it differently. There is the shape of the frame like a ghost in the grass, the stalks around it bent and flattened. The bees are dipping out from the hive, down to the place where the honey was. To find it isn't.

'They must be getting sent out to it,' he says, building the theory as he goes. 'What's it called, a waggle dance?'

I know about the honeybee waggle dance from another

book I've been reading, *The Buzz About Bees*, by Jürgen Tautz. The 'dance' is used to recruit workers to a particular nesting or foraging site, and is remarkable in that, through a series of precisely choreographed movements, sufficient information is passed between workers to enable them to find their way to a specific location even when they have not visited it previously.

A bee 'dances' by anchoring herself to the comb with all six feet and rapidly 'waggling' her abdomen, then turning in a full circle and waggling again, before turning in the opposite direction to mark a figure of eight. Other foragers gather around her, following these movements with their antennae. The angle of her body represents the direction of the pollen or nectar source in relation to the sun, and the duration of the dance indicates the distance to travel. She dances more energetically if the pollen or nectar source is particularly sweet; less so if there are obstacles or threats to be negotiated en route. Honeybees don't learn to waggle dance; the dance is innate. When reared in isolation, they're able to perform and interpret the movements even after a first foraging flight.

Laurence shifts his weight beside me. Is it a waggle dance at work here? It's possible, but I don't think so. Dances become vaguer when objects are near at hand; no longer dealing with objects remote in time or space, the dancer would only be able to communicate that a pollen source is '*very close!*'.

'Maybe some honey dripped when I picked up the frame,' I say. 'Perhaps they're smelling it in the grass.'

One, two, three more small bodies dip out of the hive, probing the flattened grass before lifting up and back inside. Gradually the numbers nudging the ghost-frame dwindle, and we look up to see that the foragers tipping out of the hive are instead turning up, away, over the beech hedge and into Hannah's garden.

'They're re-orientating,' Laurence says, and laughs, and I feel a loosening around my shoulders. It might be hopefulness about the hive or it might be something else: relief that a creature – any creature – can change direction, start over, stop feeding at an empty frame. We pull ourselves up and walk back to the house where there are berry stains on the bottoms of pudding bowls, and on mouths and spoons. Dulcie has abandoned her chair and is sitting on the floor with her back pushed up against the radiator. We spread out. Gather cushions. Later we'll eat again, and then they'll play music. For now we sit and chat. Our voices are lively and attentive and they soften the room, soften my edges.

A few times that afternoon I hear myself talking about the situation at work. The second time it sounds repetitious and the third time it sounds stuck. I wonder if they're asking themselves why I'm still there, why I haven't quit. 'I'm trying to adapt,' I explain, to myself or to the room. 'I'm trying to learn some *perseverance*.' It sounds a bit odd when I hear it spoken aloud like that; a bit high-minded and unnecessarily obscure.

'Some environments are toxic,' Dulcie says bluntly. 'They have to be changed.'

Once more in the early evening I glance out at the hive from the window, and then I see journeys lifting like invisible threads from it.

On Sunday there is half an hour just after my friends have gone when the house is empty but still full of them, and the rooms are quiet but still ringing with them. I fill the sink with pots and pans and go into the garden. Sit down on the stump and see that patch of grass at the base of the hive where the ghost-frame was, the stalks straightened now.

I am still learning how to *look*. And when I think back over the last few weeks, I have a hunch that, so far, I've been foisting a lot of my own anxieties onto the bees. Wanting them to stay, to be okay, to show me that a wildness can dwell here. Those don't sound like great terms for a relationship, or for understanding what's *really* happening in the hive.

So how to care, without caring too much? I don't want to detach completely, but I do want to recognise the bees for the separately alive creatures that they are. I recall again that letter of William Mewe's about the bees being changed by his looking. I'm not sure about the bees, but my friends do change me and what I'm capable of noticing. My own eyes are not always the most reliable ones and sometimes I need other people around to help me see, or to see differently.

I pull my suit on, zip up. Stand at the hive and grip, lift, listen for that hummed disturbance coming from inside. Slowly, I begin lifting the bars up. I'm still half-thinking about the music last night, and the lunch we ate today, and the washing-up soaking in the sink; so when the comb appears it is suddenly, weightlessly. A single piece attached to one bar near the middle of the hive, where it hadn't been before. Fresh, almost white. Almost translucent, when I hold it to the light. No twist.

5

*Losing Sight*

# May

The weather is getting warmer. Spring has been busy unfastening itself, and the meadows around Oxford's city centre are filling with weeds and wildflowers.

In the garden, the hive is thickening. I sit on the stump most days, and like this I have learned some things. The bees are busiest just after noon. You don't disturb them if you're sitting quietly, but sometimes they come in for a closer look. They are more interested in faces than feet. It's possible to tell the difference between a worker and a drone bee from a distance of two metres – you just have to look out for the size of their abdomen and for their eyes, which are differently spaced.

At the hive they are bodies; at the fence they are black and determined specks marking flight paths in the air.

The workers are out searching for pollen and nectar, but the drones aren't. By this time of year they have another reason to leave the hive, and it has nothing to do with foraging or with swarming away.

Reopening Bee Wilson's *The Hive*, I learn that the question of honeybee reproduction was a source of debate and contention for centuries. In ancient Greece, some thinkers proposed that bees don't reproduce at all but instead find their young nestled inside plants and flower heads. Aristotle wasn't convinced, but he did consider that sexual intercourse must be out of the question for honeybees, since he'd been unable to identify the male and female sexes within the hive. Workers couldn't be female, he reasoned, since they have stings and *nature does not assign defensive weapons to any female*. But – busy as they were with housekeeping and raising young larvae in the hive – nor could they be male, since *no male creatures make a habit of taking trouble over their young*. The bees were slipping between categories again, and – unable to fit them neatly – Aristotle concluded that the workers were neither one sex nor the other but a mixture of the two.

I'm beginning to enjoy this slipperiness of honeybees, their tendency to evade categories, which had even Aristotle confused. The mystery surrounding their reproduction persisted until the time of François Huber, that blind natural historian of the eighteenth century. And now in the evenings I find myself returning to his letters, following his investigations into the impregnation of the queen. The bees in my garden seem tentatively to have established themselves in the hive, but I've been wondering about this question of honeybee reproduction. Since beekeeping is about keeping something alive, not just fixing it in place, I want to find out how a colony regenerates itself – where does its new life come from?

There are plenty of references to Huber on the internet but only three portraits of him that I've found. The first shows a young boy with fat cheeks and heavy curls and an inquisitive expression on his face. He's looking up, as though to an adult, or anyway someone wiser. It reminds me of old-fashioned Christmas cards and cherubim children and I find it a bit twee.

The second portrait is of Huber as a grown man. His hair is unruly and the top button of his waistcoat is undone despite the formal sitting pose. He looks active and robust but his eyebrows are big and his eyes are oversized, as though the artist has fallen prey to that habit we have when we're trying to ignore something, of instead focusing in, of being unable to look away.

In the third portrait, Huber is old and gaunt, and this time the artist has painted his blindness for all to see. His mouth is not quite closed, his eyes are sitting slightly askew. He appears troubled. He gazes wildly into the middle distance with the air of someone unaccustomed to holding a focus. I don't like this one very much; it makes Huber look deranged. It doesn't fit with the picture that's formed in my own head from reading the letters he wrote.

François Huber was born in Geneva in July 1750 to a family of intellectuals. The family was wealthy and there was no expectation that the young Huber should earn a living. Attending college in Geneva, he was nevertheless a high achiever and studied hard, often working and reading late into the night. By the age of fifteen this intense schedule had begun to take its toll. He contracted an illness that left him weak and seemed to be affecting his vision. He was taken to see the physician Théodore Tronchin, who prescribed rural respite without delay. Sent to a village just outside Paris, Huber apparently recovered his strength, but his eyesight continued to deteriorate. His father arranged for him to visit the renowned oculist Baron Michael de Wenzel who, curiously, gave each eye a separate diagnosis. In one eye he found cataracts (untreatable at the time) and in the other *gutta serena*, a term used to describe blindness where the source of damage was located in the posterior region of the eye. *Gutta* is a Latin word for drop, and tended to be attributed to diseases of that time which produced a fluid discharge. *Serena* implied a clear and undisturbed exterior. Where no scar or inflammation was visible, the cause was understood to be behind the eye – on the retina, or in the optic nerve, or the brain. Huber's blindness was pronounced incurable.

By this time Huber had met and fallen in love with Marie-Aimée Lullin, daughter of one of the syndics of the Swiss Republic, who did not approve of the match. Despite her father's protestations and Huber's own fears over the kind of life he could offer her given his failing vision, the two of them were married. As well as his wife she became his reader and also his secretary, often helping him in his work.

Huber said of Marie that *as long as she lived I was not sensible of being blind*, and in their home he could live as if he wasn't. All the paths in the grounds around their house were hung with knotted ropes so that he could keep up his walks in the open air unaided. The cord was a thing to hold on to, and the knots were markers to help him orientate himself.

At a time when traditional modes of seeing and knowing were being replaced by a new spirit of scientific empiricism and rational enquiry, the honeybee must have made a compelling subject for study. The true origins of all substances associated with honey-making – brood, wax, pollen, nectar, propolis – were still unknown, and dispute persisted over the act of reproduction itself. Male and female sexes had by now been identified within the hive, and it was understood that the queen, not the workers, laid eggs inside the wax cells; but the means by which those eggs became fertilised had not been established. No one had yet witnessed a mating between queen and drone, and in fact the Dutch biologist Jan Swammerdam argued that, given the proportions of the male and female sexual organs, intercourse was physically impossible. Theory and speculation abounded. Perhaps it is not through intercourse but something *else* that sperm is passed, posited Swammerdam, noting that drones seemed to emit a strong odour at certain times. Or what if bees are like frogs and fish and fertilisation comes *after* the eggs are laid, suggested the English naturalist John Debraw. Or, argued Johann Hattorf, who came from a region in Central Europe known as Lusatia, perhaps the queen impregnates *herself*.

I notice, as I put Huber's letters down for the weekend, that the house is drying out. Finding myself operating some selectivity over which particular wildnesses I want to live alongside, I've scrubbed the mould patches from the walls and they haven't come back. Yesterday I left the back door open and then the wind swung in, chilling the surfaces, unsettling anything light and not fastened down.

Down at the hive a lot is happening. I only open it up once a week, so I can't keep tabs on everything, and have instead begun following hunches, looking for clues. Why did they build this way, why not that? Where has brood been laid, how far does the comb reach?

Some weeks when I open the hive the bees fuss around me like a bad temper; others they are quiet and slow and I can move the whole way through the hive, lifting each bar in turn, without eliciting much more than a grumble from them. I can never tell before I lift the lid which way it's going to go.

There is something about this act of moving through the hive from one end to another as they cluster, rise, foam, that gives me the sensation of working always very close to an edge. I am never more than a finger's width from a commotion; never more than a moment from outstaying my welcome here. If I flap, they panic. Forget gaining mastery over the colony; the task will be to learn a bit of self-mastery.

At the base of the hive is a removable tray which sits underneath the wire-mesh floor and collects any detritus slipping through. Once a week I take it out, lay it down on the grass and sift through the wax- and pollen-scrapings and pieces of dismembered wing. I'm looking for *varroa* – a tiny red parasite that over the last few decades has become common in hives worldwide. A healthy colony of bees will manage the problem for themselves by picking the mites off each other, but it's a good idea to keep an eye on whether numbers are increasing or decreasing, because it helps gauge the health of the hive as a whole.

This week I count twelve mites – enough to fit on a single fingertip. The bees zip back and forth above my head, in and out of the hive entrance. There's a brittleness about those armour-plated outsides; something crisp and crackable. I try not to bother them. Sometimes as I replace the roof or the underfloor tray I squish one.


During the week, in the evenings, I return to Huber's investigations into the impregnation of the queen.

Aided by his assistant François Burnens, who had arrived at the house as a peasant servant, Huber began repeating and adjusting the experiments of his contemporaries, testing their conclusions against his own. Always he seems to opt for the most meticulous, and so the most arduous, approach. In one experiment he had Burnens catch, hold and carefully examine every single bee in two separate colonies; it took a total of eleven days. Often they hit dead ends or apparent contradictions and found themselves confused, and none of this is omitted from the letters. *I have no regret*, he wrote. *By going over the same observations several times, I am much more certain of having avoided error.* We are unable, of course, to ask Burnens if he had any regrets.

I'm fascinated by this perpetual struggle for knowledge; the lengths Huber, and so also Burnens, went to in a bid to solve the mysteries of the hive. It is an investigation into honeybees but it is also an experiment in wanting; in willing a breakthrough.

Swammerdam had proposed that the queen was impregnated by way of an odour secreted by drones, and Huber tested this by confining all of the drones in a hive inside a tin box pierced with holes, separating them from the queen while allowing scent to permeate. Like this she remained sterile; she could not be impregnated by smell alone.

The French scientist René de Réaumur had also been studying the queen. Wanting to catch a glimpse of the act of copulation among honeybees, he'd tried putting a virgin queen and a number of drones inside a glass jar and waiting. Huber and Burnens tried this too, experimenting with drones at different ages over different seasons, but – like Réaumur – they saw no act of copulation take place. *We detected between them a sort of union, but so short and imperfect that it was not likely that it had operated fecundation.*

On discovering a whitish fluid at the base of the wax cells, Debraw had become convinced that eggs were fertilised in the same way as frogs and fishes, by the male injecting a seminal fluid *after* they were hatched. Huber and Burnens also observed this whiteness at the cell base, but found it to be an illusion *caused by the reflection of ... light*, where fragments of cocoons, successively hatched, accumulate.

By this time Huber was almost completely blind. All observations came to him through Burnens, and the vivid descriptions in his letters must have resulted from a combination of Burnens's ability to watch and vocalise, and Huber's own propensity for questioning and cross-checking. He must have been listening very carefully, too. From the collected impressions and remarks of a sighted observer a picture was forming in his mind of the hive's interior. The picture would prove a more coherent and more accurate conception of the bees' world than had ever before been described.

Huber had worked his way through each of the existing theories of impregnation, but none of them seemed to lead anywhere. Then one day he and Burnens tried a different tack. They stepped back, this time pushing all hypothesis and supposition aside and instead returning to what facts were available to them. What were the *core conditions* necessary for impregnation?

They removed all the drones from a colony and narrowed the hive entrance to prevent the queen from leaving, or any more drones from entering. Like this, she continued to lay eggs that hatched successfully. Now a new thought occurred to Huber. In this and previous experiments, researchers had used queens whose prior history could not be ascertained. Was it possible that mating had happened *prior* to her confinement in the hive? They repeated the experiment, this time with a queen that had been isolated since birth, and found that she remained sterile. No eggs were laid. But here they hit upon a surprising finding. As a control measure they'd confined another virgin queen inside a hive *with* drones present, and here the same result occurred. It appeared that a virgin queen remained sterile when confined, *whether or not* there were drones present. Perhaps the answer to the question of impregnation lay *outside* the hive.

The next weekend I am visited by my friend from Leeds. His name is Dan and there are blue wisps in his dark-brown hair from when he dyed it for a party last month. He stands by the shed watching as I open the hive, arms folded nervously across his chest. 'You just carry on and do your thing, love,' he said as I put on the suit, stepping back. He shrieked when I put the gauze mask down, but he is quiet as I lift the lid.

Inside they're growing. It's not the bees themselves that are growing but the colony itself, and the comb which is lengthening from the top bars towards the base. At first the new comb was so pale it seemed almost translucent and I worried there were no eggs inside it; when brood has been capped with wax it's supposed to be the colour of digestive biscuits, but this was more like shortbread. It was all so new – there were no layers, and nothing soiling them. A couple of weeks on, the comb is thicker. It's toughened into yellow, and almost brown in places. I look back at Dan standing by the shed. He's craning his neck without moving a step closer. I can see his socks. They're bright blue, like his hair wisps. How is it that he is always so perfectly matched?

In the middle, near the centre of the comb, I find the queen. Her head and upper body are the same size as a worker's, but her abdomen is huge; round and shining like the smooth outside of a chestnut bud. She rotates slightly, her backside nudging as though it were the one doing the directing, and not her head. I hold the comb higher, squinting, and her attendant workers gather in a clutch – mounting her, hiding her, wary of my gaze and the glare of the sunlight.

The hive has often been touted as a symbol for the virtues of a monarchical society, but that's misleading. In fact it is the workers, not the queen, who hold much of the decision-making power; and in the absence of a ruling authority it is communication, not control, that maintains the stability of the hive. Which rings true with the sense I have, standing here, leaning over them like this, of something alert and meshlike at work.

'That was *wild*!' Dan says, when the lid is closed and the suit is off and I'm standing by the shed with him. But I'm not sure *wild* is how it felt. My eyes are wide and my heart is beating faster, but I also feel calmed and – well – sort of contained, I suppose. There is something quietening about looking in on the bees. I have to steady my legs; concentrate on what I'm holding and not on their seething; find a way of picking out what's happening beneath all that heat and constant motion. It makes me feel a bit forceful myself.

'How are they doing?' Dan asks, tweaking my elbow, nudging me back down to earth.

'They're getting bigger – I saw the queen.' I pull him closer to the entrance until he sees their differently dressed-up legs. There are pollen buds the colour of traffic cones and dumper trucks, and some are even an odd grey-green.

'Looks like mould,' he says, eyeing it suspiciously. 'Where does it come from?'

'Not sure. All over, I guess. Allotments. Parks. There's a meadow just beyond the playing fields, so maybe there.'

Three more come down, carrying that odd grey-green. He's right; it does look like mould.

Next he wants to know about the queen. 'There were so many of them,' he says. 'How does it — who do they — how do they know where to find her?'

I search around in my head, trying to find an answer for him. 'Well,' I begin, 'they can't *see* her. It's crowded, and pitch black in there. So she sends out a pheromone — a particular smell, like a signal to let them know she's in attendance. It gets licked off by the workers around her, and they pass it on to the workers around them, and so on. They spread it body to body through the whole colony.'

Workers also use smell to communicate. After a rainstorm last week I saw three of them standing over the entrance, waving their backsides at the freshly watered air. They were releasing a different pheromone to guide home any foragers who had become lost. And they use scent to recruit foragers *out* of the hive, too — sharing samples of nectar during a waggle dance. In fact, if you think of smell as a form of communication, then even the dead speak: the last odour to be emitted by a bee is oleic acid, the smell of decay. On smelling this an undertaker bee will seek out the corpse and begin dragging it towards the entrance. After some distance she will drop the load and move on, and another undertaker will arrive and drag it further, and then another, until the entrance is reached and the dead bee ejected from the hive. I saw this too last week. I was sitting watching the entrance when suddenly something dropped. I thought at first it was a stone falling, but then half of it moved. It was two bees wrapped around each other, and one of them was alive. The live one began tugging. The dead one was so tightly curled I couldn't see its legs. It got tugged almost to the edge of the concrete slab before the alive one left it.

We head back up the garden, sit down on the doorstep with two chipped coffee mugs and the bottle of wine Dan brought. We're still within sight of the hive, and looking over I wonder what's now being said, what messages are being passed inside it. I like the thought of a stability that comes from fine-tuned communication, and not the say-so of a single ruler. It must be a restless kind of stability, I think. The messages come constantly and from all around, and catching them is more about receptivity than reach.

'So I've been meaning to ask you,' Dan says, leaning back against the doorframe, settling himself in, 'are bee-keepers celibate?'

'Oh,' I say. Crossing my legs, recrossing them. 'I don't know. Why?'

'Well, I was chatting to this guy online,' he says, and his eyes shine as I raise my eyebrows at him. 'He's a *farmer*. And he was telling me he's been dating this fifty-nine-year-old beekeeper.' He sips his wine and pauses. I see that he's enjoying this. 'But the thing was, the beekeeper couldn't have *sex*. The guy was celibate. Something to do with the bees, the farmer said.'

I am not sure what to say about celibacy and the hive; I've never heard of that tradition. The thought makes me uncomfortable, and I've been getting comfortable here, and I wish he hadn't mentioned it, all casual and nonchalant and from the blue like that. So I change the subject; I tell him about Huber's letters and about the mystery of the impregnated queen. I'm getting ready to begin the story of Huber's loss of sight and how much he nevertheless understood, when Dan reaches over and ruffles my hair.

'Been staying in a lot recently, have you?' he laughs. 'You're turning bee-obsessed. Why don't you give yourself a break? Go away for a weekend, let your hair down. Have some fun, for Christ's sake! You can't be a beekeeper all the time.'

I'm caught off guard; I'd expected him to be fascinated by my story of the impregnated queen.

'Maybe,' I tell him, thinking probably not. I shouldn't leave the bees on their own, I think. Not yet. We were just getting used to each other. What if something happens, and they need me here?

I've been telling myself that the rather closed routine I've slipped into lately is for the bees' benefit, but in fact I'm getting used to it. It feels important; necessary, even.

The hive has become a counterweight to a work environment I've been finding stressful. I've begun to relax out here; to drop some of the rigid outer casings that were holding me stiffly and rather unhappily in place. The bees are hot and busy inside the hive, and maybe if I can understand them better I might learn something important about how to live.

Thinking this, I feel a flash of kinship with François Huber; his withdrawal to a place where he could reconceptualise the hive, create a world afresh. There's a loneliness about a situation like that; it makes for a lot of disjunction when the world beyond your fence doesn't match up with the one you've built inside your head. Huber wanted to demystify the colony, and so change beekeeping practices everywhere; I imagine the intense detail in his letters must have been due in part to the criticism he anticipated from among his peers. It's a difficult kind of longing to carry around, wanting the world to be otherwise – to bend its shape, let something new in, accommodate you; and it can trap just as surely as it can set free.

On Sunday Dan packs his bag, changes his socks, and heads for the train station. I try catching up on chores in the house but that challenge of his to go out, *have some fun*, has bothered me.

I remember that a friend in Islington wants someone to look after her kittens when she and her partner go away in a few weeks. And then it occurs to me that I could go back to the British Library for more of Huber's letters. So maybe a trip to London is not such a bad idea after all.

I call Luke and ask what he thinks about the idea. Can I leave the bees alone for a weekend? Would it make me a bad keeper? He laughs and says they've done okay without me for millions of years; they'll be fine for a few days now. Feeling a little foolish, I call my friend with the kittens and ask if she's still looking for someone to come and stay. She is, so then it's settled. I book a train ticket. And since my bee books are sitting right there beside me I pick one up and flick through the index, just to see what it has to say about celibacy and the hive.

Worker bees remain celibate throughout their lives. The queen releases a pheromone that suppresses their ovaries, so although they spend a lot of time caring for young brood they don't actually lay eggs and never become sexually active. In ancient Rome, Virgil invited his readers to marvel at this selfless virginity, but he didn't suggest they copy it. We're humans, not bees, and creatures of the flesh; having sex and giving birth are all part of being human.

Reading this, I have the rather childlike sensation of having been found out. I haven't been paying much attention to human flesh recently; I've been busy paying attention to the bees. Perhaps I've got caught up again, gone *too* close – my caretaking has tipped over into iden- tification. I imagine Virgil wouldn't think much of my impulse to absorb a few life lessons from the hive.

The book I've picked up is *The Hive* by Bee Wilson, and now I skim through the pages, searching for traditions linking the sexual activity of beekeepers with the celibacy of bees. Wondering if there are stories of mimicry, or blending, between keepers and their colonies.

The Greek essayist Plutarch claimed that bees were especially bad-tempered towards men who'd recently had sex, and that they could sense adultery and punished it by stinging. The Roman writer Columella advised beekeep- ers to *abstain from sexual relations* the night before opening a hive, and the Italian Renaissance writer Rucellai claimed that bees could identify an unchaste person by the smell of his breath, which repulsed them. In a tradition common across eastern Europe, a girl's virginity could be tested by having her walk past a beehive: if the bees left her alone, her purity was judged to be intact.

They're curious tales, not about identification with honeybees so much as compliance with social mores, and more about friction than blending. I'm not keen on the moral undertones but I do enjoy the switching-around of roles between the keeper and the kept. In these stories the bees are the ones doing the seeing, gaining access to our hidden places; the keepers are the ones being seen.

Standing by the hive, in the presence of a colony, it is difficult *not* to notice that very special receptivity among the bees, and I wonder if it was partly this that led thinkers past to suppose that bees can sense the things we'd prefer to keep hidden. Looking in on a colony is a bit like watching a weathervane trembling and flitting in a wind; you can't always pinpoint what the bees are reacting to, what messages they're carrying. It is just like our species to assume that the messages are for us, and concern ourselves.

In June 1788 Huber and Burnens, still deep into their investigations of the impregnation of the queen, made a breakthrough. Knowing that drones tend to leave the hive at the warmest part of the day, just before noon they seated themselves in front of a hive in which a queen had been confined since birth. *The sun had been shining from its rising, the air was very warm; the drones began to fly from several hives.* Enlarging the hive entrance, Huber and Burnens paid close attention to the departing bees. Sure enough, the males soon appeared and took flight. Not long after this the young queen emerged. She paced back and forth across the entrance, lifted her hind legs to her belly, and pressed. Stood and faced the wind. Suddenly she took off, rising twelve or fifteen feet in a series of concentric circles. Huber and Burnens sat below and looked up, but soon she was out of sight.

After only seven minutes she was back again. This had been her first experience of flight, and outwardly she appeared unchanged. But fifteen minutes later she re-emerged from the hive, tracing the concentric circles higher and faster so that they *soon lost sight of her*. Almost half an hour passed before she returned, this time very different from before. Her sexual organs were *filled with the same substance [which] very much resembled the liquid contained in the [sexual organs] of the males*. As she disappeared inside the hive this time Huber and Burnens confined her, and after two days found her belly swollen and a hundred eggs laid inside the comb.

The evidence appeared unequivocal: queens mate with drones outside the hive. Huber was now able to account for the high prevalence of drone bees in a colony: since *the queen is compelled to fly in the expanse of the air to find a male, [it is] requisite that these males be in sufficient number for the queen to have the chance of meeting one.*

Huber was right to conclude that queens mate with drones. We know now that on sunny afternoons all the drones in an area will leave their hives and gather in the air at a specific location known as a *drone-congregation zone*, where all the virgin queens in the vicinity will follow. Reaching this crowd of flying males, a queen will secrete the pheromone decanoic acid, which attracts drones by signalling to them her sexual readiness. High in the air she mates again and again, up to twenty times in a single afternoon, often returning to the same place over several days. Afterwards she'll remain fertile for the rest of her life, laying up to two thousand eggs a day in summer – more than her own body weight. Unless the colony swarms, she won't leave the hive again. This will have been her sole experience of flight. Each time she mates with a drone, the drone dies because his penis and abdomen get ripped out.

It's true that I've been hiving away, ensconced in the letters of an eighteenth-century natural historian or else tucked down by the far fence; but I'm also beginning to feel like I'm *getting* somewhere.

I'd noticed a curious tic in Huber's letters – a gap or a slip in language that I kept getting stuck on. When in 1792 he published the collected letters, it was with the title *New Observations on the Natural History of Bees* – but why that word 'observations' if, in truth, he couldn't see?

It wasn't a one-off; rereading his letters, I realised that they also contain frequent references to vision. *We did not believe our eyes*, he wrote. *I saw . . . I observed . . . Before our eyes . . . Under our eyes . . . I have often seen.* I assumed at first it must be down to faulty translation, but when I checked the original version in French I found the same: Huber consistently referred to *seeing* when describing his experiments with honeybees.

The published volume begins with a preface acknowledging his blindness and the debt he owed to Burnens – *[I can] see well when seeing through his eyes* – so he wasn't seeking to conceal it. Far from clumsiness or faulty translation, those slips signify something important. Through that process of intense learning and investigation, perhaps observation came to represent something else to Huber, other than vision: a capacity born out of relationship, built on trust and the slow, careful extraction of detail from the contents of experience. Like this, *observation* could encompass smell or touch, close listening, an unfaltering attention; and it might proceed through a spirit of constant questioning, imaginative leaps and patience.

The next time I open the hive, it buzzes at me. The sound is like a friction, and I realise that I've been marked as an intruder. Soft bodies harden and throw themselves at my chest. These are guard bees, warning me away. Perhaps they want to shake me up a bit, stop me getting too comfortable here. They're asserting their separateness; stating their right to this place. I find that I'm glad of it. That the bees are defensive must mean there's something in there worth defending, and as I lift the first bar I feel a new weight to it. The colony has grown again. The comb is stronger-looking, and deeper than before. I turn it one way and then the other, knotting my elbows so as to keep it straight and level with my face. It's clustered with bees. I can hardly see the comb, there are so many of them. I go through the bars one by one, noting in my head where there is fresh brood, pollen stores, first signs of honey-making. And all the time there is this soft low buzzing that thickens as I lift the bars and settles as I replace them.

I don't see the queen this time. They're getting too crowded; it is difficult to tell head from abdomen.


The doorbell doesn't work, and the door rattles when Dulcie knocks on it. She's come from London, my friend with the pudding-bowl hair and the clothes like a dressing-up box; she's staying here a night while she attends a study course just down the road.

It's dark when she arrives but she wants to see the hive so we go out. Without suits we crouch down and press our ears against its sides, straining through the traffic and the Friday-night street noise. We can't hear buzzing, but Dulcie thinks she hears a growling coming from inside.

I imagine a murmur. Tens of thousands of attentions pricked at our footsteps and brushing cheeks, as if listening could make its own kind of sound.

'I bet they can hear us,' she says. The earth is dry. The air is cold. She turns her head and whispers something into the hive, but the hive doesn't make a sound.

At the time of Huber's writing, most scientists believed that honeybees were deaf. Unable to find ears or an auditory system like our own, they had assumed that bees were without hearing. But this posed a problem. How to explain the complexity of processes that went on inside the hive if bees were without ears and couldn't see in the dark? Did they communicate by smell alone? And – if they couldn't hear – why would they be making sounds? Not just buzzing but also hissing, quacking, tooting, piping, warbling, beeping; as measuring instruments became more advanced, more noises were added to the list. The scientists scratched their heads, and the bees tooted and hissed.

Then, just as with Huber's investigations into the impregnation of the queen, one day a breakthrough came. The scientists lowered their eyes, away from the honeybee's head, to her feet. This is where she listens; sound enters through the floor – through contact with *things*. Inside the hive she stands on the surface of the comb, and the comb shakes. Elsewhere, workers shake their abdomens against the comb; she senses this as a pulse rising through her feet and knees, and it signals to her that the colony is preparing to swarm. A shorter pulse will prompt a dancer to stop dancing; a similar movement on the body of a queen will inhibit laying. These vibrations, which are passed at the resonant frequency of the comb itself, can be heard by a honeybee despite the noisiness of the hive, and at a distance – the comb amplifies the sound. Detected by subgenual organs in each of her knees, the vibrations are converted into a nerve impulse and sent directly to her brain.

Next the scientists looked up again, to her antennae. Here she feels a back and forth. While human ears respond to changes in air pressure, her antennae detect air-particle oscillations. A wing beats, the air moves and she hears it. Sensitive to frequencies of up to 500 Hz, her antennae appear to have adapted to discern the 200–300 Hz frequencies produced by a dancing nestmate as she vibrates her flight muscles nearby. While the existence of these communication pathways is well recorded, the precise patterns of signal and response, and so also their functions inside the hive, are still not fully understood.

I may not be undergoing a transformation into a honeybee, but I like learning things like this because it stretches my range of the possible.

Dulcie shivers. We stand a moment and look up at the

orange blur of light pollution seeping upwards from the horizon, thinning and bluing, soaking the night.

'Not unbeautiful, is it?' she says.

Our feet scuff at the too-dry earth. The garden seems narrower in the night: the trees over the fence blend with the warehouse roof so that they are more looming than in the day, and the traffic seems closer too.

There are times, not often, when a wall crumbles, a breakthrough comes, in a place you'd thought it wouldn't. Sometimes this happens with a big announcement and a noisy fanfare but more often it comes quietly, as though slipping in through a side door, by way of no more than a pause or a shift in the conversation.

'You should meet my friend,' Dulcie says, glancing back at the hive as we turn and head for the house. 'He's a beekeeper too. Maybe we can all meet up when you're in London next week. You two have just the same taste in furniture.'

I laugh and say that I didn't know I had any taste in furniture but yes, okay, I will meet this beekeeper with the chairs that are just like mine. And a few seconds later we're back in the house and the lights are on, it is our own faces we're seeing reflected in the windows, and we are already talking about something else.

*

In the end, when the weekend of the Islington kitten-sit arrives, I am glad to get out. I go straight to the station after work and catch the train to London, then sit in my seat and watch as the view outside the window begins to crawl, and then to flow past. Soon everything is green and billowing.

It's last-minute but anyway I begin texting friends to invite them over for a meal on Sunday night. I have the number of Dulcie's beekeeping friend because she's been trying to arrange a drink between the three of us, and without thinking about it too much I add his name to the list in the 'to' box and press 'send'.

And then each of my friends, for different reasons, can't make it. Dulcie can't get a babysitter; Laurence has plans already. The third ate an old tin of tuna and has terrible food poisoning, and the fourth is on her way to Glasgow. The fifth and sixth say can we meet up tomorrow, it'd be much easier tomorrow. So then it is just Dulcie's friend the beekeeper left. And he replies to say that's kind, yes please, he knows that area well. He used to live just around the corner and he'll be working not far from there on Sunday; it'd be very easy to come by after. I wonder if it is a little strange to invite a man I have never met before to a house I have never been in before, and I hope that he doesn't mind pasta.

*

The friends who rent this house have just moved over from America so the place is sparse, and simple in the way that two graphic designers can make a place feel simple. The handles of all the mugs on the kitchen shelf are turned to forty-five degrees. The books are unbitten at the edges, and they have more pictures in them than words. The mantelpiece in the front room is the only place showing any sign of disorder – it's crowded with toy cars and a plastic lion and pieces of bark and string, like the flotsam a bird might find and take to nest with.

I spend the weekend cycling around and seeing friends, playing with the kittens or else extricating myself from their claws. And then it is Sunday, and there I am waiting for Dulcie's friend the beekeeper, hoping we'll find something to talk about or that, if we don't, he'll leave quickly.

His hair is almost completely white. This is the first thing I notice when he arrives at the door, and it's a surprise because he is only just about my age or a bit older. It changed colour when he was a teenager, he'll tell me later. Something switched, and all the follicles in his head lost their colour. 'Did something happen?' I'll ask, before thinking to censor myself. 'I've wondered that too,' he'll say, cocking his head, sweeping his fringe to one side. His voice is quiet and sometimes I'll strain to hear it.

But that will happen later. Right now he's standing in the doorway; he has to duck his head to get in.

'Dulcie told me you're a beekeeper,' I say, stepping back to let him through.

'I'm not a beekeeper,' he says as he takes off his shoes, 'I don't know where she got that idea.' And then there is a pause. 'My uncle's a beekeeper,' he offers. 'Oh dear.'

I make tea as the kittens roll and scramble around our feet. There's a slight crease to his frame that makes me think of lifting, not hunching, as if the bones in his shoulders are gathered close under his skin. It makes me think of the brittle chitin of bee wings, of those veins inside running right to the edges of them.

We eat roasted squash stirred up with some spinach and pasta and a few other things that I find in the fridge. It is not especially nice but it is a bit nice, with the green leaves and the bright-orange squash all mixed up on top of each other. Sometimes I think colour is as important as taste, I tell him, then wonder if that sounded pretentious or overly self-assured. He says he has a friend who eats different-coloured foods on different days, depending on her mood. Red foods on angry days, green on feeling-alive days, purple on sad days. Only there aren't many purple foods, he says, so she doesn't eat much on those days. Mostly just beetroot. I know someone who does the same thing with clothes, I tell him. I like beetroot, I say. He says so does he. His accent is Yorkshire, I think.

After the pasta is over we go into the room with the cluttered mantelpiece and I sit down on the floor instead of the sofa. Like this, I am at the same level as the kittens, which feels like a good place to be. He looks a bit confused, then sits down on the floor too. The kittens tug our jumpers with their claws.

'Last autumn I did a course about being a psychic,' he says, then looks at me, surprised. 'I haven't told that to any of my friends.'

I narrow my eyes at him. He doesn't look much like a psychic. 'So can you read my mind?' I joke, because I'm not sure I believe in psychic ability, and feel I should make that clear right away. But what he then describes is not so much about reading minds as developing a different kind of attention to things.

'I'm not very articulate,' he says, and stops. His eyes are very blue. 'Let me show you.'

I have to stand with my eyes closed and my palms out in front of me, and try paying attention to what's happening around me in the room. He's going to stand at the opposite wall and begin walking slowly towards me; I have to open my eyes when I can 'feel' him. It's a bit like the children's game Grandmother's Footsteps, except there's no one around to shout '*Wolf!*'.

'Sometimes you find you can sense a person even before your hands touch,' he says. But the first thing I notice when I close my eyes is my own anxiety at the thought of something proximal and approaching.

'I can't do it,' I say, and open them again. He hasn't moved an inch. 'I can't tell the difference between my own nerves and what's coming from around me.'

He grins. 'Exactly. It's about learning to discriminate.'

I notice he's taking this seriously. I close my eyes again. Maybe, nearly, my palms feel something. And then suddenly we remember the curtains are open, and there are passers-by watching us from the street. Two people standing at opposite ends of a room, hands up like buffers at either end of a train track. And then two people doubled over laughing.

That night I dream there are holes in my feet. The holes don't hurt; they are just *there*, newly opened and completely strange. I feel air moving beyond, inside the place my skin had sealed. In the dream I reach down, touch a hole, poke my finger in and twist. My right shoulder jumps and my ear twinges and a shiver runs right down the back of me.

The next morning I get up and leave early. At the station I hit rush hour, a thousand suits, announcements booming over loudspeakers and upraised faces as departure boards flash with destinations and times. I feel strangely vulnerable; I keep thinking I'm going to get knocked down. And my train is about to leave. I duck between umbrellas and luggage trolleys, slip through the barriers and squeeze into a carriage just as the doors are closing. The aisle is full of bags and bodies and elbows poking. Grey suits apologise and huff. We retrieve our screens and papers, lower our eyes, tuck our necks back inside our coat collars. The train pulls out of the station and I gather myself, take out my wad of photocopied letters from the British Library. By the time we've reached London's outskirts I am already back in the eighteenth century.

Where Huber's experiments began in a spirit of fascination and excited enquiry, they later take on a stranger tone. He and Burnens move in on the bees themselves, mutilating bodies, isolating them from their sensory capacities and observing the effects this has.

It is all in the name of science. There is nothing particularly unusual about his methods; you learn the function of a thing by noticing what changes when it's gone. Huber has Burnens remove wings, pick an antenna off. Paint varnish over the eyes of workers to blind them. And, through Burnens, he begins to observe the effects this has – on the bees, on their behaviour and capabilities, and on the capabilities of the colony.

The descriptions in his letters are as vivid as ever, and alive with compassion and feeling. But in one experiment on a confined and mutilated queen his language slips again, and she seems to be accorded a whole personality, an inner life of her own. He has amputated her antennae. Without them she is unable to feel her way into cells for laying, and instead *drop[s] her eggs haphazard[ly], without thinking* (they scatter and fall, producing nothing). She shows *symptoms of delirium*. Unable to fit her eggs securely, she flees, seeking out the empty parts of the hive, its hidden corners, where she becomes motionless. A number of workers *follow her in this solitude* but she avoids them, rarely reaches for food, and, when she does, sends out her proboscis *with an uncertain groping* towards the heads and limbs of those in her vicinity.

Isolated from her sensory world, she becomes *tormented with the desire of leaving*. She rushes at the opening, and – finding it blocked – returns to the hive, indifferent to the care of those around her. When Huber and Burnens open the entrance she rushes again, and tries to fly, but – her belly swollen with unlaid eggs – she falls. The workers do not follow her. She dies.

I struggle to locate what it is I find so moving about this last description of the confined and mutilated queen.

That, trapped, she panics; that the workers attempt to save her, to feed her; that she refuses to be fed. That this situation becomes unbearable; that her only option is to flee.

Huber wonders what it is that causes the workers to abandon her at this point, as she falls from the hive, but can draw no firm conclusions. *Her instinct is changed*, he writes, *this is all that I perceive. I see nothing more.*

The following week I learn about a contemporary version of Huber's experiment happening right on my very own doorstep.

On Wednesday I meet Jack for a walk after work. He's been busy. He's got so interested in beekeeping, he's started building hives with a group at a local community allotment. Every now and then he's been coming to open the hive with me, and he wants to set another date for that too. And then there is a meeting to plan a summer music festival, and a gathering of the outdoor-swimming club at the weekend. He's planning a trip this summer; he's going to swim the length of the River Dart. 'Of course, I'll have to walk some of it,' he adds. 'Some of it is just a stream. Anyway, how are you?'

I can't seem to shake the feeling this week that something's missing. But I don't tell that to Jack – it sounds too strange – so instead I tell him about the bees, that pollen the colour of grey-green mould that they're still carrying in through the entrance.

'That'll be the raspberry flowers,' he says. And then he gets out his phone and shows me a website he's found, with colour charts to help identify pollen sources. 'You match up the pollen colour to the chart, see? And then you get an idea of what they're taking into the hive.' We stop in a shop doorway to peer down at the palm-sized screen. And this is how I discover that just as I have been exploring further, so the bees have also been flying further, moving into more places than I'd thought. They've been inside hawthorn bushes and up horse-chestnut trees; somewhere or other they've even found a patch of raspberries.

Jack has recently joined a campaign to stop the use of neonicotinoids, a class of insecticide that first came into use in the UK around twenty years ago. Neonicotinoids work systemically. Unlike contact pesticides, which remain on the surface of treated foliage, neonicotinoids are taken up and transported to all parts of the plant, remaining active for many weeks and so removing the need for repeated treatments. You can look at a field of maize or a garden of flowers treated with neonics and you won't notice anything unusual. Like *gutta serena*, that disease affecting one of Huber's eyes, there is no sign of damage and no visible trace of anything altered. It's the same with the insects neonics act upon. Working from the inside out, they penetrate the brains of insects, blocking receptor sites and exciting nerves, leading to eventual paralysis and death.

'They were initially hailed as a safer alternative to other treatments,' he says, as we slip off the pavement and down onto the canal towpath. 'The pesticides would work discreetly, the manufacturers said. From the insides of plants, directly into the brains of insects – so they'd stay separate from the soil and water supply.'

A moorhen peeps. A duck lowers herself into the water, and five ducklings follow her.

'It didn't really work out that way,' he says, and goes on to tell me how more and more evidence is coming to light to suggest that the effects are far more wide-ranging, and more dangerous, than was first thought. With their highly evolved nervous systems, honeybees are especially vulnerable to neonics; studies are finding that some of their key abilities are seriously impaired following exposure to the systemic pesticides, even at the relatively small doses that make it into the hive.

'Like which abilities?' I ask, kicking a pebble and watching as it trips over the gravel and plops into the grimy water.

'Well, like communication,' he says. 'Homing and foraging ability. The way they learn. One study showed that neonics affected bees' ability to tell which scents signified a ready food source. They were losing the ability to connect smell with meaning; that amazing system they have for perceiving and making sense of the world was starting to fail.'

There's a new development of modern flats on the bank opposite, built in a semi-hexagon, so that all the windows face in towards each other, with a paved car park at its centre. I suppose the idea was to offer residents a canal-side location, but from here it looks more like a panopticon.

'Honeybees aren't solitary individuals,' Jack says, as we watch a man drive into the car park through a security gate with a keypad that he pushes a code into. 'Their ability to function as part of a social group is crucial. And when something's getting blocked – when the capacity of individuals to communicate and learn is being impaired – well, it impacts the whole colony. Lots of colonies. However many colonies are living across a whole area.'

It sounds like one great big experiment into the severance and confusion of the senses, and it makes me sad, and I wonder what will happen to the bees.

We've walked as far as the edge of the city, which doesn't take long really, and now we stand and look at what comes after it. There are electricity pylons and fields. A tractor must be moving along somewhere, because there's a soft plume of dust rising, a spray of gulls circling above it.

'Shall we walk a bit further?'

'Okay.'

A narrowboat plugs by, with a man in a tracksuit and a woolly hat standing at the helm. He nods, smiles and raises his mug in our direction. The mug is painted with bright blue flowers.

'A drone-congregation zone,' he says, and laughs easily. I'm surprised at my relief. 'Have you ever seen one?'

I'm back in London. My friends with the kittens are away again, and the psychic has come over for another visit.

'No,' I say, remembering how I stood in the garden this week, scanning the sky for one. 'You'd probably need special equipment. Apparently there's one over Crystal Palace, but I don't know. Mostly they're a mystery.'

'A mystery?' he says. 'Hm. And what about your queen? Has she visited one of these congregation zones?' I'm not sure if he's playing with me.

'Oh yes,' I say. In fact, my queen is laying furiously. There's more brood in the hive each week – the workers must be struggling to keep up with her.

'I've brought wine,' he says, as though he's just remembered the bag that he brought in with him.

'Oh—' I say, 'I don't have a corkscrew.' I notice he's

watching me, carefully. 'Maybe we can borrow one from the neighbours.'

There's a block of flats opposite, and I've already crossed the road before I remember I am not wearing any socks or shoes. I wince. Turn and look back at the house, where his face disappears from a window.

A man stops in the doorway to the flats and begins searching his pockets for keys. 'Excuse me,' I say, and he looks nervously at my feet. 'I'm sorry,' I say, holding the wine bottle up, 'do you have a corkscrew?' He looks at the bottle, then back at my feet. He is probably not asked for corkscrews very often.

'Okay,' he says. And then, hesitatingly, 'I live on the top floor.' So I stand on the doormat and wait as he goes up to fetch it. My toe pricks, and I pick a shard of glass from it.

Back in the house I am victorious. 'I forgot my shoes!' I say, brandishing the opened bottle.

'I watched you from the window,' he says, and it sounds a bit like a confession. 'When you looked up I stepped back and nearly trod on a kitten.' There's a spot of blood on my toe and my feet are still tingling.

There is something about him that is difficult to place. This is what I think to myself as I turn and take two glasses from the shelf. And then I realise: he holds himself still. We cook, and talk, and drift between the sofa and the cooker, and all the time there is something in him that sits. Like this, I notice how much he sees of things; and how constantly I move around.

And, like this, a space opens between us. It's a live space, alert and curious, like a tension held. He shows me how to cook the rice just right (whatever you do, don't stir it), and he neither invades nor retreats. We sit on the floor and chat as the kittens scuff the carpet around our knees. (It is not until later, when he's gone home and I'm falling asleep, that I will realise I might want to touch him.)

I walk him to the Tube. It's busy outside, full of Saturday night lights, and our strides get bigger and our grins wider because of it. We turn off the main road down a narrow side alley to take the long way round, then suddenly he stops.

'Look!'

The moon.

It is full, and huge. So low that it might have just now descended and settled itself there at the far end of the street, between a church spire and a chimney pot. It is like some great lamp or orb or an eye beaming. There are other people around us on the pavement. They are also stopping and pointing and looking up.

6

*Swarm*

# June


There's a cloud of them thronging the entrance. It's just after midday, and the weather is warm; a kind of traffic jam has formed outside the hive, with dozens of bees hovering and circling. More appear at the hive mouth, readying themselves for flight. There are a lot of them. Probably this is not the best time to make a hive inspection, but anyway that is what I'm doing. I pull on my suit, zip up. There really are a lot of them.

I lift the roof and begin easing the bars out. Guard bees fly up, pronging my chest and head. With each bar that lifts I make a fresh hole in their ceiling and they spill out, crawling up the sides and over the tops of the wooden slats, onto my gloves and wrists. I can't feel them on me through the suit, but I can feel how the air changes because of them. It thrums and bristles and stiffens and shakes, and all the while I have to keep steady and slow and my movements must be soft – I mustn't let my attention or hand slip. It seems to work best if I stand close but slightly apart from the hive, trying to keep some sense of the whole as I pause over each part of it.

It's looking good. The comb has formed in perfect parallels, and the bars lift smoothly up, and nothing sticks. It's all down to that bee space – the exact distance the bees leave open between each comb, to allow air and movement through the colony.

I want to see it. I prise two bars apart, widening the gap enough to peer inside. Between the combs it's dark and thick and clodded with bees, and their wings rustle as I draw closer. The bee space measures three-eighths of an inch. That's the length of a fingernail, or half the width of my eye – it is a small space, even for a bee. Just enough for one body to move alongside another. And yet the whole life of the colony passes through it.

'Okay, are they?' Becky asks as I arrive at the back door with my suit still on.

'I think so. The hive's getting very full.'

'You're covered!' she says, stepping back. She points to my shoulder, my head, both knees. I brush them off, one by one, and step into the kitchen.

*

I've seen the psychic again. Really I should stop calling him the psychic, but I find that I can't say his name. He's a musician; he makes music. I've heard it, and it's really good.

We sit on a log in a park in south London eating toasted sandwiches sticky with melted cheese, and talk about being kids and growing up and finding a voice.

'I think that's why my music is so quiet,' he says. 'My mum didn't approve, and the walls were thin. There was only a very small gap for it to exist inside.' I poke my finger into the plasticky grey of my takeaway cup and fish the teabag out. I'm thinking about bee hearing, those fine-tuned channels through which signals run back and forth; and about bottlenecks and one-way valves and things being released and hidden. I flick my hand and a spray of tea lands on my shirt and on his jumper. When I look up I see his face, hill, open sky, and an oak tree far off in the distance.

'I like your buttons,' he says, pointing to my jacket. 'Those are good buttons.' The buttons are metal and I sewed them on after I'd pulled all the old ones off. They're so shiny that if you put your nose up close they become like mirrors, and then you can see your face and what's behind you in them. I don't tell him this because I'm not sure what it would mean to invite his nose up close.

As I walk away after meeting him each time I have a feeling of being differently weighted. I am quieter as I move around, and lighter, like something emptied. I notice the soles of my feet. The soles of my feet notice the ground under them. Not just firmness or softness but also movement, like the rumble of this lorry passing.

He's gone back to Forest Hill, and I've got as far as the bus stop. My phone rings and I jump, even though the ringing is not so loud and definitely quieter than the traffic on the road beside me. It's Luke. He's calling to arrange a visit. We've talked a lot since the bees arrived, but he hasn't been back to see them.

'How about next week?' he says. He's speaking from his flat in Soho, where there's glitter in the bathroom and painted nudes on the walls, and Walkers crisps and books and pieces of hive stacked up in the kitchen. 'I could get a train to Oxford one afternoon, and we could open the hive together.'

'Great!' I say, relieved at the thought of having him there to act as an interpreter. 'Come for dinner!'

The bus arrives, and I get on it.


'I got dive-bombed!' Becky says when I get home that night. 'I was watering the lettuces, and one dive-bombed me!'

'You want to be careful,' I tell her. But this is not

helpful, since there is only so far that you can go with being careful when there's a colony of bees in your back garden.

You can't stop a colony from growing once it's started – it will just keep on wanting to expand. And so much goes on in those journeys beyond the garden, it's not like you can do anything to moderate their intake. The bees will keep going out and bringing back, and as I watch them thickening around the entrance I begin hoping the hive is enough to hold them.

I am itching to see Luke and for him to see the hive. On the day of his visit I rush through work and leave early; he's already waiting when I get home, standing outside the front door in red trousers and a shirt patterned with tiny sunflowers. '*Helen!*' he beams as I get off my bike, offering tea or coffee or whatever. 'Later,' he says, waving a hand. 'Let's see those bees.'

Inside, he drops his bag and we pull on our suits, but I want to stand back and watch today so when we go out to the garden it is he who opens the hive, not me.

'Look at that,' he says, lifting a bar out. 'Now that's a good start.'

The comb attached to it is thick and full; at least as long as my forearm. The walls of a single hexagonal cell are only eight-hundredths of a millimetre thick. Honeycomb is rigid, and feather light when empty, but a piece like this from a top-bar hive will weigh as much as three kilos when full.

It is easier to look when someone else is doing the lifting. So with Luke holding the comb I peer in, trying to make out individual cells under those thousands of crowding feet. Each cell is built with a slight upwards tilt so as to keep whatever is inside from falling out. Inside we spot eggs and larvae, soft and mucuslike, surrounded by pollen and nectar and a soft rim of honey stores. But we have to be careful. If Luke wasn't holding it just right from the bar at the top the whole thing might rip and tear under its own weight.

A comb-building bee must form a cocoon of warmth around herself as she works, to keep the wax soft and malleable. She secretes wax from glands in her abdomen. It comes out in flakes and she moulds it with her mandibles, which are like two shovels on either side of her mouth. Mandibles are for cutting (she can slice the leg or tear the wing from an opponent), as well as for chewing and manipulating. She uses them for filling holes with propolis, and for shaping the wax cells.

A bee lands on my hand, and I lift her up to see the mandibles. Everything about her is somewhere between reaching and receiving; she probes the world, tastes and touches it, as she bites and chews it.

Luke is unhurried by all their fuss and teeming. He replaces the first bar and levers another out – this one is just as full. To hold comb like this you have to heave, and it takes effort to keep it lifted. You feel the muscles in your shoulders tug, and there is always a pull in them for a while after.

'Well,' he says, when we're nearly through, 'there must be nearly a hundred thousand bees in here. It's the biggest colony I've seen all year.'

It's big. I repeat this to myself. The colony is strong; and the hive is nearing capacity. There are only three or four empty bars left at the far end of the hive. If they expand much more, it will be full.

'It's unlikely,' Luke says, when we're back in the kitchen and pulling our suits off, 'but you might find they swarm before the season's out.'

When a hive is full, the colony will split in two. The queen's pheromone that usually gets passed around the hive no longer spreads far enough; there are too many bodies, and the bees are building too far from her. Without her smell around to act as a suppressor, the workers will begin raising a new queen. Once she's hatched, half the colony will leave with the old queen in tow to found a new hive elsewhere.

I imagine a swarm leaving – only half of them left, the hive emptied and thinned. With that open doorway at the hive entrance you might be doing everything right and still there's no guarantee the bees will stay. Then why put all this effort into *keeping* them here, making them strong, if they'll only end up leaving?

Luke has folded his suit and is placing it in his bag; he turns and takes it out into the hall.

'But,' I say, wanting to stop him, backtrack. 'But what can I do?'

'Do?' he says, half-listening over one shoulder as he hangs the bag from a coat hook.

'How can I stop them swarming?'

'*Stop* them?' he says, hands still raised with the bag in them.

I stare at him, incredulous. *Yes, stop them.*

With the colony diminished the hive would be more vulnerable to attack, and the remaining bees would make less honey too; but the feeling in me is simpler than that. I want to keep them here, inside *my* hive. I've grown used to that curious hiving activity tucked away by the far fence — attached to it, even. I don't want to see the colony depleted and I don't want any of them leaving.

Luke ducks his head through the doorway as he comes back into the kitchen. 'Swarming is a colony's repro-ductive function,' he says, 'it's a natural process. It's how they make more of themselves.'

'Of course, you might want to *contain* them,' he adds, filling the kettle now, searching the cupboards for mugs. 'You don't want a load of bees flying loose around the city.'

Thoughts are rearranging themselves, positions are shifting in my head. I am trying to keep up. So you can think about the reproduction of the bees themselves, or you can think about the colony as a whole – with needs and drives and an impulse to procreate, to venture out into new areas and outposts. It comes back to that in-betweenness of honeybees, their wild uncontainability. That the colony has a reproductive impulse seems obvious now I know about it; but somehow I hadn't figured it until now.

I open the fridge and pull out a bag of carrots, trying to suppress my own impulse to go outside right now and clamp the hive closed, stop them moving around for just one bloody moment so I can sit and get the measure of this.

'Tea?' Luke says, passing me a mug. I take it despondently and begin making soup, as Luke explains about swarms. How you might stop one, if you wanted to. Or contain it.

It's possible to watch a colony for signs that swarming is imminent, he tells me. You might notice queen cups among the brood – protruding, globular cells that look like they've distended and drooped – a sure sign that the workers are preparing to raise a queen. You might also notice the workers behaving differently. They'll be resting more, and foraging less. They might be eating honey rather than making it, and raising less brood than usual.

'You can try stopping them at that point,' he says, 'by making more space inside the hive, or by looking out for the queen cups and destroying them before they hatch.'

'And does it work?'

'Well, temporarily. If you can keep them contained until the flowering season's finished, when the colony will reduce naturally again. Or you can create an artificial swarm. Beekeepers who don't want to lose their bees often split the colony themselves, taking the queen out and putting her into a new hive along with half the colony. That way they end up with two colonies instead of one.'

We eat sitting at the table, with bread and cheese and the tail end of a bottle of wine. There are grains in the bread that catch in my teeth. Chunks of half-submerged carrot parsnip potato float like islands in our bowls, and we fish them with our soup spoons.

'You know,' Luke says, leaning back in his chair, 'next time you're out there, you should widen that entrance. You won't stop them swarming by keeping it tight like that. You'll just piss them off. Try opening it up a bit – just keep an eye out for wasps raiding.'

Luke is good at letting things be as they are. He told me about wax moths once. They can destroy a whole hive in one go, eating through the comb until there is nothing left. 'What a disaster,' I said. But he said that even wax moths have their place, if you think about the hive itself as a kind of ecosystem. In the wild they move into abandoned hive sites and clear them out, removing old and dirty wax and reopening cavities ready for new settlers to move in. 'Like a decongestant,' I said, imagining a forest full of bunged-up trees. And he said, 'Yes, I suppose so.'


When the soup is finished Luke gathers his things and heads back to the train station. It's still light outside when he's gone. I open the door to the garden, go back out to the hive.

Over by the fence I flop down on the tree stump, wipe a fleck of soup from my sleeve. And wonder about the seeming futility of a role that tasks one with keeping a creature in a place without holding it down. It's like a hopeless riddle. In fact if the hive is confirming anything it's that things move and change no matter what you do, so maybe this whole exercise in beekeeping is really a kind of farce and I should just give it up right now and save myself some hassle.

I hunch forward and huff, elbows on knees, chin in hand–heels, my back to the house. I am unsettled like the bees get when you start lifting the bars and rearranging the hive – they're already trying to resettle themselves, to regain control, before the rearranging has even finished.

If I'm honest, giving up wouldn't feel quite right. Not really. The bees might not need me but I want them here, and want to be able to care for things without fearing they'll lift away. So how to care in an open-handed way – *with the hands, and hence with attention*, as the dictionary once suggested?

The grass around the stump is wet and long and when I tug it comes up in a big clod, roots and dirt and all. In the soil a worm appears, sun-starved and shining, and begins working his way back down.

I think of Luke coming all this way just for a hive-inspection; how well he looks after his friends, no matter the distance. Perhaps it's what makes him a good bee-keeper, too. He treats bees as he treats people, as wild, alive and curious things that move and grow and fly and change as the seasons change around them. I don't always find that easy; but if the choice is between detachment and the more unpredictable, knotty business of attachment – which lays us open to pain and loss and faulty translation, but also to joy and journeying and fresh meetings? Well, then I know which I choose.

The last few foragers are returning for the evening and now I lean in and widen the entrance like Luke said. Inside the house I spend a lot of time worrying about the bees, but mostly what I *do* out here amounts to not much more than this. Opening a space a little wider in one place; sealing it off in another. Perhaps in truth it is the hive, and not the bees, that I am keeping here.

A *hivekeeper*.

I turn the word over in my head as I walk back up to the house.

It has a good ring to it.

The week after Luke's visit it seems like there are swarms everywhere. I get four phone calls in as many days from people wanting to tell me about ones they've seen. My parents arrive home to find one in their garden; Dulcie glimpses one in the doorway of a shop on Oxford Street. Another friend calls me from Brighton where she's hurried in from the street, which a swarm of bees is filling. She's standing at the window with her baby under one arm, holding the phone to her cheek. 'What's it like?' I ask, imagining something like you get in old cartoons, those black dots all moving in one motion. 'They're *everywhere*,' she says, and her voice is high like a thing unhinged. 'They're all over the place!'

A friend in Hackney sends a picture of a swarm hanging like a crust over a car bumper. Another clinging to a lamppost. The pictures are uncanny – a wild thing taking on the forms of features in an urban landscape.

I want to see one for myself. I begin looking especially carefully at cars and lampposts.

On a lunch break I call Ellie and tell her about the crowdedness in my hive and about all the swarms flying around everywhere, and she says maybe I'm in need of a bit of light relief. After work I cycle over to her flat and we sit out on her fire escape, where she's growing geraniums in pots, and drink elderflower cordial mixed with fizzy water.

She already knows about the Islington kitten-sitting and the psychic and the oversized moon, so now she wants to know what else.

I come clean. 'I've been listening to his music,' I tell her. 'A lot.' So much, in fact, that I've begun wondering if he's noticed a spike in listeners on his website. 'Do you think it's possible to trace online hits? Find out who's listening?'

'I'm not sure,' she says, 'I don't think so.' Ellie is not a lot, but a bit, more computer-savvy than me, so I decide that probably she is right, and am reassured.

She wants to know about the bees too.

'How are they doing?' she asks. 'Any signs of swarming?'

'I don't think so. But it's getting difficult to see. There are so many of them. Silly, isn't it? I spent all that time worrying the colony was weak, and now here I am worrying it's too strong.' I snap the stalk of a geranium flower, pull a deadhead free.

'That reminds me,' she says, 'I have something for you.' She goes inside and comes back with a bowl of olives and a notebook, sits down and rests her feet against the railings. 'I was thinking about swarming. Remember when I mentioned the verb *hive*? How the meanings seem to contradict each other, so that in one sense it's about collecting and gathering together, but in another it's about breaking away, making separate? Well, it occurred to me this week that the verb *swarm* is just the same.' She leafs through the notebook until she finds the page she's looking for. 'Here we are – I made a list. To swarm is *to gather . . . to come together . . . to assemble . . . to crowd*. But it's also *to escape from the parent organism . . . to found a new colony*. Funny, isn't it? Like two opposing pulls in one same motion. I wonder if your bees are feeling a bit torn.'

Perhaps they are, but it isn't the bees that her list has made me think of. Don't those two opposing pulls also describe the feeling *I* have as I open the hive? That sense of escape, of touching the edge of something foreign and new; and also the feeling – as I look – of gathering focus and attention? As though I retreat to the hive so as to come together; escape in order to find some reassemblage. Which is a kind of homing, I suppose.

From the top of Ellie's fire escape you can see what's happening in the driveways and back yards all down the street. Next door a man is building a large wooden structure in a parking lot. Behind him is a garage hung with a sign saying *XMAS TREES*, but there are no trees for sale at the moment. He doesn't notice us watching him, or if he does he pretends not to.

We spit the olive pits into the patch of grassy rubble at the bottom of the steps where there are dandelions growing, and nettles and cow parsley and a small clump of ox-eye daisies.

'I wonder what it feels like to be a colony preparing to swarm.'

'Well, you'd be feeling constrained, I suppose.'

'And confused.'

'And wanting to take flight.'

'But uncertain about leaving, maybe. It's risky, no? To fly up without knowing where to.'

'You'd be watching the light at the hive mouth.'

'Mm, and then a rush of flight. All those bodies throwing themselves out.'

Fifty thousand bees exposed and freshly vulnerable,

Swarm

flung upon the air. I wonder if I've been experiencing something like a swarming impulse recently; pondering a boy with white hair and blue eyes who lives a long way from here, wondering what a step into the unknown might do.

'But how do they do it?' Ellie asks.

'Do what?'

'Leave, all in a mass like that? How does it happen?'

I realise I'm not sure.


When I get home that evening I prop the scrap of note-book paper with Ellie's copied-down definition on my bedside shelf, beside the piece of torn-off comb and the bottle of hornet whisky. Then I search for my button jar.

It's at the bottom of a cupboard, underneath a pile of shoes. I pull it out and turn it around in my hands, sending the buttons tinkling. Some of the buttons I've collected myself, others I've been given. Mostly they're different colours and shapes, but if you empty the jar and spread them out it is possible to find a few that are the same. This is what I do. I sift through them all until I've picked out six of the metal ones with the shiny mirror faces.

Next I drop the buttons inside an envelope, and write the psychic's address on the front. This looks a little formal. So then I take a magazine and flick the pages until I find a picture of two painted animal masks. I cut these out and stick them to the envelope, beside his address. At least it is a bit more colourful now.

On the way to work the next day I pop the envelope in a post-box, but when I hear the papery swish of its landing inside I wish suddenly that I hadn't. I wonder about hanging around until the postman comes, asking if he'd mind returning it to me. Probably he'd be too busy. I check the collection time: six o'clock this evening. It is a long time to wait.

At work I cover my desk in paperwork, make lists. File things into cabinets. As I'm eating lunch at my desk I knock a glass over and watch in horror as the water floods, the ink blossoms and the paperwork ripples and rucks and turns a light shade of blue. I should pull myself together, I think, mopping up. It is no good to go around spilling things. Thank goodness it's Friday, and there's a weekend ahead.


With the hive entrance widened the bees are bottlenecking less. There's less congestion and more flow to their passages back and forth, but inside they're seething. I've begun scanning for signs of dormancy – some indication that there might be a swarm brewing. There are no queen cells and the bees are busy, but when I opened the hive this week I found another bar full – they're still expanding fast.

I've been trying to keep steady, not do anything yet except wait and keep watching for the signs. But, away from the hive, I've wanted to know more. I'd like to understand how a swarm forms; what happens inside the hive to trigger a colony to split. What is this movement that is about both gathering and breaking away? On Saturday I'm restless, unable to settle to anything. So I fetch my books and spread them out.

There's a story we like to tell ourselves about the hive as a place of cohesion, with all bees working towards a common goal. It's not true. In *The Honey Bee* by James and Carol Gould, I learn that what actually happens is a lot more unsettled and shifting. The decision to swarm doesn't arrive as a single event; it's a much slower and less defined process, whereby queen cups (the protruding cells in the comb in which young larvae are raised as queens) are continually being built and taken away. This is true even in a colony with a healthy queen and no particular impetus to swarm. One worker might begin preparing a queen cup on an area of brood comb, and the next might tear it apart. Or the next might add to it, and then another, so that the queen cup begins forming, until a fourth worker comes past and destroys it. Later another worker begins repairing the cell, and another. Eventually a threshold is reached; a new queen is raised, and the colony swarms.

This ambivalence of the colony – whereby questions remain open, and answers are always in a state of process – is highly adaptive. The bees respond faster and more flexibly to changes in their environment than if all acted with a single mind.

I sit up, and close the books. Human movement, also, is tempered by this push-pull. I lift my arm and feel the muscles contract and stretch. Imagine the bees out there, inside the hive, sensing thresholds. That constant building and tearing away, in a place between staying and leaving. The solid wooden frame and the sturdiness of its legs; all those thousands of wings rasping. The sound of them, when I open the lid. It is almost like paper rustling.

There's a lesson in here somewhere. As I feel new spaces forming, new possibilities opening beyond the hive, I too have been preparing to lift up, break out. I feel ready for it. The bees have chewed through some of my congested bits just like the wax moths do; I'm feeling better resourced, more in touch with things around me, more able to begin something new. Perhaps in a way – unbeknownst to them – it is the bees who are the open-handed ones; they who are setting *me* free.

'Helen!' Becky's shouting to me from the kitchen.

'Something's happening. They're buzzing. Come quick!'

I join her at the door. The smooth flight paths have disappeared and the bees are churning around the entrance. There's a loud humming, almost a throbbing, coming from the hive.

'Perhaps they're about to swarm,' Becky says, and I nod, but they don't, and a while later they are quiet again. No swarm has departed, and the colony hasn't split. But something has happened to unsettle them.

The first picture he sends is of that brown envelope that I put in the post-box, taken on his phone. It has landed on the edge of his doormat, one corner and an animal mask perched over bare floorboards.

The hive is like a heart and it pulsates. Or the hive is like a brain with a million synapses. I can't decide. Anyway, there's more honey in the hive this week – and still they haven't swarmed.

There are occasions when bad weather or some other event external to the hive can mean that a swarm is imminent but delayed. At such times there may be a few moments in which two queens – one unhatched, the other waiting to depart – communicate through the comb using sounds known as *tooting* and *quacking*. The old queen *toots* by pressing her thorax against the comb and sending out a vibration, causing the workers to freeze. If the unhatched queen is old enough to respond, a *quack* will be felt through the comb. On hearing this, the workers will move to keep her forcibly in her cell until the old queen has departed with the swarm. Once a virgin queen has hatched, she'll use the same system to locate and kill any other unhatched queens by stinging them through the comb; if two or more queens have hatched they'll usually fight to the death, and the survivor will replace the old queen.

The next time I see the psychic I visit him at his flat.

'You're wearing *dungarees*?' Becky says, as I'm about to leave.

'You think better not?' I say, glancing over at her astonishment.

'Not unless you want to give him the impression that you're a kid. A kids' TV presenter, or a *kid*.' I am not sure what impression I want to give, but anyway I go back upstairs and change into a pair of trousers the colour of a terracotta pot and a black vest, and I plait my hair and put a pair of sandals on and then I come back down again.

'*Sex!*' she says, when she sees me. 'Great!' Which makes me nervous.

We meet at Forest Hill station and he grins when he sees me, and springs a little on his feet. 'I've just finished work,' he says. 'I could do with some air. Shall we go for a walk?'

The Horniman Museum is at the top of his road and there are gardens in the grounds where you can sit down on the grass, so we buy a bottle of beer from the petrol station around the corner and walk in through the gates as the sky turns pink behind the distant skyscrapers and cranes. It is easy to feel as though you can see the whole of London from here, but I imagine London is probably bigger than it looks.

He's going on holiday to Scarborough next week, he says. I've been to Scarborough, I tell him. I ate fish and chips. And he tells me about a haunted house on a hill nearby and about the sea and the old fishermen's cottages along the coast.

The conversation wanders on, and I notice that I feel quite bare beside him, as though when I speak I lose my coverings. I realise it right then and there, which is unusual, since normally I don't notice things until it's too late and they're already gone and finished.

Whoever is in charge of the Horniman gardens has decided that they should not stay open indefinitely. Just after sunset a man with a golf buggy and a big handbell arrives. He clangs the bell and walks around a bit, and then he clangs again. This is the signal that you should pack up your bags and go, so we get up, but the man is not paying us any attention. He's busy arguing with two girls at the bottom of the slope who won't get up and don't want to leave and why doesn't he just fuck off. He clangs the bell in their faces and gets back in his golf buggy, and as he drives off the two girls pull themselves up and head for the exit gate anyway. A fox standing behind a tree a few metres from them has been watching all this unfold.

We walk back to his flat. Outside his bedroom window is a box bursting with geraniums, bright red like the ones on Ellie's fire escape. There's a bird feeder too, but it's empty.

'They get through it in a few days,' he says, seeing me looking. 'I have trouble keeping up with them.' I give him an understanding nod like I know how it is with wild creatures, you can never be enough for them; just wait until I tell you what's been happening with the bees.

But I don't tell him about the bees just then. Instead, I tell him I'd like to kiss him. And then watch his surprise. There is a pause of a few moments that feels like a very long time indeed. He is not uncomfortable. I am immensely uncomfortable. This makes him smile. His nose comes close up.

And the time to leave and catch my bus arrives, and I don't leave. Later, I have a feeling of something pulling free, somewhere in the middle of it, as if my ribs are splitting open. It is a kind of violence wrought through tenderness, and it comes as a shock.

'This doesn't happen to me very often,' he says after.

'This doesn't happen to me very often either.' The morning is still early when I leave, and the air is cold enough to sting my teeth. I am peeled, and ringing.

'Like being steamrollered,' Dulcie says matter-of-factly, after I've blundered up a hill and down a road and in through her front door. She's talking about falling in love. We're sitting in her kitchen and I can't talk much because I might have lost the place where my language comes from. Or the place where my language comes from might have shifted, or something. Her kitchen is white and clean and I'm watching her move around in it.

'What, like emotionally?' I say. 'Like being overwhelmed?'

'No,' she shrugs. 'Just like being steamrollered. All my organs mashed into a pulp.' She turns and takes a tin of sweetcorn from the cupboard. She's got twenty minutes while Corinne naps to clear the surfaces and have a shower and get everything organised before college. She's back in full-time study this year, training to be a speech-and-language therapist – juggling essays and exams alongside and in among everything else. I can't seem to move; I'm just sitting there watching her do everything. She lights the gas, empties the sweetcorn into a pan.

'Doughnuts!' Her partner, Neil, appears in the door-way, rubbing his eyes and yawning; he works nights and has just woken up, ready to take his turn as childminder. Dulcie looks at him and grins.

The doughnuts are sitting in the middle of the table, still crispy and warm, greasing the waxed bag that I brought them in. I saw them in a bakery window on my way here, and ducked in to buy us one each. We all sit at the table now and eat them slowly, the grease smoothing the tips of our fingers as the sugar sticks to them and to the edges of our mouths.

We don't lick our lips.


The second picture he sends is of a road sign perched high on a cliff-top road just outside Scarborough. The sign is an upside down triangle and it says Give Way. Beyond is a rim of field and the wide-open sea stretching far into the distance.

# July

I've understood that swarming is a natural function of a colony, but what about when it happens in a city centre? What do you do when thousands of bees begin flying around, or clinging to things, right there in among the cars and kids and offices and shops, and everyone trying to go around as normal? What then? That's what I want to know.

I find out at the next Oxfordshire Natural Beekeeping Group meeting, which I am hosting, when we're visited by our local swarm collector. Every urban area now has a dedicated swarm collector. This is who to call if you find a swarm, so that they can come and safely remove it for you.

The load at work has eased a bit in recent weeks; the pressure lessened. Perhaps people's attention has strayed – there is a lot of sunshine outside. Or perhaps I am a bit sunnier myself. Anyway, with more light and longer days I have more energy to hand, so it felt good to offer our house as a meeting place when Paul sent an email round with the date (*Tea would be nice*, he advised, *but no biscuits. Biscuits set a precedent*).

The swarm collector arrives at my house before anyone else. He's wearing a bright-orange builder's vest, which I expect him to take off, and he doesn't. His hair is frizzy and grey and it's tied up in a big ponytail which hangs bushily down his back. Next to arrive is Paul, whose arms are full of books and boxes, and then fifteen or twenty more, far too many to fit in the front room, so I open the back door and they filter out as I gather all the cups and mugs that I can find so as to meet everyone's tea requirements.

I spot Helle, Jude and Mary, but there's no sign of Mark, who has to drive twenty miles to get here. There are a few long skirts and quite a bit of flowing fabric, a lot more hair, some office suits and an anorak. They're a funny bunch, and I like them. Each one trying to keep his or her own corner of wild nature in the city. One lady, recently divorced, will be keeping a hive for the first time this year on a rooftop near the city centre; another has two hives and four children and lives in a small cul-de-sac in a village just out of town; she's got a rabbit hutch too.

I move around the group with a pot of tea, filling cups. Inside the pot, fresh peppermint leaves are stewing and tipping against the china sides. When I pour it, the water comes out softly green.

Paul gathers us around so he can introduce the swarm collector, whose name is Mo. Then Mo takes centre stage. He gets around sixty calls a day in high summer, he tells us, though most of these won't actually be about honeybees. 'People get confused,' he says. 'Mostly they're looking at wasps or bumblebees.'

The group forms a circle around Mo, who is leaning with one elbow against the hive. 'You have to really push them,' he tells us. 'People tell you what they think you want to hear, just so you'll come. I have to get them to describe exactly how high the swarm is from the ground. If it's too high up or inaccessible for whatever reason, there's no point in me going.' He's wearing big boots and he's speaking louder than anyone has ever spoken beside my hive; I imagine the bees inside hushing. 'And,' he goes on, 'truth is, usually they're miles away. So they'll tell me, "Ah, it's just a metre off the ground!" But really they're inside the house with all the doors and windows closed, and the swarm is in the neighbour's garden.' He stamps a foot, prods a steel-capped toe into the grass. 'People can't judge distances,' he says, looking around at us. 'That's what I've found.'

'If they're on a low branch, or something removable, it's easy. You just cut the branch they're clinging to. Hold a bucket underneath, and they drop right in. Sometimes I don't even wear a suit,' he says, his chest puffing under the luminous vest.

After Mo's talk we break out into smaller groups, and I take a seat beside Paul. I tell him I've been learning a lot about swarms this year.

'It's a hot topic,' he says, and shuffles excitedly in his seat. 'Many beekeepers these days order from professional breeders. Queens and even whole colonies get flown across continents – that's one of the main reasons why varroa has become so widespread. But if we allowed colonies to swarm naturally, in a carefully managed way, we'd have local stocks readily available.'

This is something that the ONBG have been organising locally for the last few years, and now Paul takes a piece of paper and a pen from his pocket and writes *Swarm List* across the top. 'Excuse me,' he says, then moves off and begins working his way around the group, writing down the names of anyone on the look-out for a colony. Over summer, the ONBG functions as an unofficial swarm-collecting network. Staying vigilant as to the movements of their bees, when a colony swarms the members alert Paul and a text message is sent around to everyone on the list and arrangements are made for someone to go and collect it. Last year everyone on the list at the beginning of the season had a new colony by the end of it.

Clearing up later that evening, I begin wondering if there's a difference between a honeybee colony that's been allowed to follow its swarming impulse, and one that hasn't. So when the washing up is done and the house is settled again I write an email to Paul, asking him, and he responds almost immediately with a carefully worded reply. He describes definite survival traits developing among colonies that have been shared through the swarm list. *Maybe it's genetic*, he writes, *maybe it's our low-intervention approach, or hive type, or maybe it's just luck. But our group seem to have bees that can tolerate and coexist with the varroa parasite without treatment, and maybe some diseases too.*

Which suggests that changing our keeping practices *affects* the colonies in our care, even when that change is about *less* intervention, not more; about stepping back, giving way to the bees' own logic and rhythms. I think I could become an advocate for the frameless top-bar hive. That odd, unlikely, trough-like shape standing out in my garden, overshadowed sometimes by the holly; tangled sometimes by the weeds. The riskiness and daring of it – of beginning from nothing, from all that empty space and readiness inside.

I wish I'd spoken to Mo more. I would have liked to ask him what disruption is caused when a swarm turns up in the middle of a city centre. And how far it can be managed, and do people or the bees get scared, and if anyone has ever been hurt in the process. But I don't have Mo's email address, so I don't ask him any of these questions, and possibly this is just as well.

Once a fresh swarm has collected on a temporary landing spot, scout bees will fly out to explore potential nesting sites in the vicinity, testing them for damp and pests and watertightness. Here begins a new process of decision-making, whereby each scout will cast a vote for her favoured location by dancing on the bodies of other bees. There are no fixed opinions, and she will visit a range of alternatives; having initially selected one site, she may switch and begin dancing for another. She takes time making her selection, and will monitor it over the course of several hours or days (the Goulds' book *The Honey Bee* describes one experiment which found that when water was poured on a particularly popular site, dances for that site dwindled). There is much at stake; the colony will be starting from scratch in its new home, and must build enough stores to last the winter. Slowly, a consensus builds. More and more scouts dance for the same location; a destination is set.

\*

There he is in London and here I am in Oxford. We are living in different cities so there are distances to travel, and gaps. Text messages in the pauses, and feelings in the wake of things. It is like Mo the swarm collector said: people can't judge distances. They get caught up, and start trying to block or bridge them.

*I wish that you were here so you could put your hand on my heart to make it quieter.*

I put the phone down and get up, draw the curtains. Perhaps I will rearrange the furniture in my room. And then my phone beeps out a reply.

*I want it to sing.*

During these first few months each new exposure is like a question: *where are we, what is this, how far might it reach?* We're testing at surfaces, searching around for a ceiling, wondering if it might be secure enough to risk building from. 'I'm going to fall in love with you,' he blurts suddenly one night, shocked by it, shocked at himself. *Can I, should I, I am.*

I see him again, and then again and again. A lot happens in a short space of time. And time does peculiar things, seeming to stretch and warp, so that we fit a lot of life inside it, as it also seems to be making life.


Back in February when we were assembling the hive I talked to Jack about Huber's blindness and Langstroth's muteness. It gave them something in common, I said. They were both missing a sense. But speech isn't a sense, he said. Senses are about reading the world, they take things in; speech is about what goes out.

I'm not so sure any more. Inside the hive there is all that fine-tuned communication happening; it is all so close, I'm not sure it's possible to tell speaking from listening, language from perception. It makes me wonder. How much of speech, how much of what we put out, is about what goes in – about how we listen, look, touch? And what does this say about the need for proximity, or for distance? I'm thinking of the detached gaze of a scientist-observer, and of the bees passing messages body to body, and of where I stand as I look at the hive and try to form an understanding of it. How close should I get, to know the detail? How far back, for some perspective?

Almost a whole beekeeping season gone, and I still don't know a right distance from them.


There's a film showing at the old cinema down our road about a family of peasant beekeepers in northern Italy, and I meet Ellie after work and go and see it. The film is called *The Wonders*. In one scene the daughter picks three bees up by their wings and places them in her mouth.

She sits very still and parts her lips a little, then the bees crawl out one by one. When they crawl out sometimes they climb up and around her cheek, and sometimes they climb down her chin.

There are traditions in cultures from across the world that link bees with speech. A ritual in ancient Egypt compared the voice of the soul to the humming of bees, and in a ceremony called the *Opening of the Mouth* the soul was released from the body with the line *The bees, giving him protection, they make him to exist*. In her book *Sweetness & Light*, Hattie Ellis describes how if bees hovered over the lips of a newborn in ancient Greece it was seen as an omen that the child would grow up with a mellifluous – literally, honey-flowing – tongue. Bees were also believed to have flown near the mouth of the infant Virgil.

After the film has finished we're unlocking our bikes from a lamppost. I've just reeled off this list of stories and ancient rituals to Ellie, and now I pause to take a breath. She puts on her bike helmet.

'But it seems a strange connection, bees and speech,' I go on. 'To be so widespread, I mean. Because bees don't actually make sounds with their *mouths* at all.' And I tell her about the places they do make sounds from: their wing muscles and their abdomens; and then there's the waggle dance and the pheromones, which are also kinds of speech.

I'm speaking too fast and I get a bit muddled, it all comes out sounding garbled and confused. And really it's late, and we're tired, and we both have work in the morning.

'It's reminding me of something,' Ellie says, getting on her bike, 'but I'll need to look it up. Check your emails in the morning.'

As I cycle home it is not the ancient rituals I am thinking of; it is that girl with a mouthful of bees. If you look very closely at something over a period of time, I imagine it must start getting inside you. Not literally, and I'm not thinking of merging exactly – but it becomes a part of you, you begin carrying it around, and I might be absorbing the bees; they might be getting into my lungs and my heart and bloodstream. Maybe one day I'll open my mouth and bees will come out, wild and humming, instead of words.

I'm making breakfast before work the next morning when Ellie's email pings through on my phone:

In Hebrew the word for bee is דבורה (*deborah*), derived from the root דבר (*debar*) meaning to pronounce or speak. In the Hebrew language all words derived from a single root are closely related, so that each derivation of debar is linked by meaning: *debar, debar, debir, deborah, midbar, midbar.* Meaning, to speak; a thing made to come about (an act, word, literary text); an inner sanctuary inside a temple (literally 'place of the Word'); bee; wilderness (a place where things live within a larger system, as words live within speech); mouth.

Speech, act, inner place. Bee and mouth and wilderness. I read through the list again. It's peculiar, the way the definitions seem to draw a neat circle around the very different themes that have been turning in my head this year. Learning how to *feel* better, and what it means to look and listen and speak; how to define *keeping* as an action, a doing, and how a colony is generated and grows; how hives and homes and heads are each a kind of inner place, each voicing something about who we are and how we relate to our wider landscapes; and how finding or losing a sense of place can feel like a wilderness sometimes.

That is a lot of things. There has been a lot turning around in my head this year.

At the bottom of the email is a short note from Ellie: *So it's a curious connection, bees and speech, if bees don't actually make sounds with their mouths. But then that fits with the word* midbar, *or 'mouth', where the 'mouth' is understood not as part of the face but as the origin or well of words.*

I find that I'm thinking again of the faceless, formless colony. Can it be said to 'speak'? Perhaps it has not one but multiple voices, which now and then combine to create sound or a decision or a movement. And the inside of the hive is where all those thousands of voices well, working over and between each other, escaping out sometimes from its ... mouth.

I drop my breakfast bowl in the sink and pick up my bag, set off for work.

If a colony can have multiple voices, I think, pulling out my bike and pushing off from the kerb, then maybe humans have multiple voices too, with different places and parts of ourselves to speak from. Not just mouths but also feet, backs of the knees, hearts. In fact maybe we are all speaking from different parts of ourselves all the time, even if we're only accustomed to listening along particular registers and lines.

We've made it through the swarming season, and the colony is still in one piece. The bees are thriving, busy outside and in. I call Luke, check in.

With the summer solstice passed, the focus inside the hive will be shifting. As the bees feel the days shortening they'll move away from colony expansion and into a more concentrated period of honey production. They'll start building stores, shoring up reserves; beginning the long journey towards winter.

'Will you take a harvest?' Luke asks.

'I'd like to,' I say. When you take honey from a hive you're also taking the bees' winter food supply; if you take more than a surplus, or if the winter is particularly long or cold, they won't have enough to survive.

'Well, wait and see,' he says. 'You'll know in a month how it's looking. And you're still opening the hive every week? You might as well stop that now. Many beekeepers won't open the hive at all through August. It'll be getting colder outside, and you don't want to chill them. Best leave them to it. Let them make their honey. Step back.'

So I do. I put my gloves and suit inside the shed, and although I keep watching the bees moving in and out, I don't open the hive for thirty-one days, from the beginning of August to the end of it.

7

*Honey*

# August

Early morning. Earlier than rush hour. Earlier than binmen or the postman or school runs or my alarm clock. The sun just lifting over the roofs of the houses opposite; the light just seeping through the curtains. The light is golden. Soon the window, the walls, my whole room, is aglow with it.

A crash from downstairs. The front door flung open, then the sound of feet and slim wheels clacking over the floorboards. Bags dumped, breath drawn. A pause.

I tiptoe along the corridor, crouch down and peer through the banisters to see Becky standing in the hall-way, back from two weeks in India.

'*Hi!*' she whispers.

'*Hi!*' I whisper back, stealing down the stairs, wrapping

my arms around her neck. She smells of airport departure lounges and of sandalwood soap. 'Welcome back.'

She wheels the suitcase down the hallway and puts the kettle on; she's pink-nosed and tanned brown by the sun.

'I haven't slept,' she says, rummaging around for her thermos. 'I won't sleep, I think. Are you up? I thought I might head out.'

'I'm awake,' I say, 'I'll come with you.'

Outside, the sun has scaled the tops of chimney pots opposite; it's already too warm for coats. We cross the road and walk away from Oxford's spires, away from the city centre, following a wire fence around the edge of the recreation ground and ducking through a patch of undergrowth.

This is the edge of the Lye Valley – an area of un-developed boggy ground (calcareous fen is the proper term, since it's fed by lime-rich springs) threaded through Oxford's east side, a hidden wash of untrammelled space.

Old drinks-cans glint up from the dirt; carrier bags shroud the trees. We startle a fox, and the fox startles two blackbirds. There's a heap of sleeping bags a few feet away that might be a person asleep under a bush.

We know about this place from Jack; he comes badger-watching here. Apparently there are a couple of setts along the bank, and if you sit quietly at dusk and keep an eye out, sometimes you get a sighting.

Becky leads the way, telling me about her trip as we go. Now she pauses a moment, scanning the slope for a path up.

'How are the bees?'

'I haven't opened the hive all month,' I tell her, 'but I will soon.'

As we climb I tell her about *hefting* the hive, an old technique of testing how much honey's inside by rocking it gently, gauging the weight. I tried this last week, and the hive did seem heavy – but since I wasn't sure how much a colony of bees should weigh, it was difficult to tell how much was added load.

'And if there's honey,' she says, 'will we take some?'

'We'll take some,' I say, 'if there's enough.'

At the top of the hill the path opens onto a golf course, the grass spongy and dense and trimmed to within an inch of its life. To our left, on the other side of this hill, are the community gardens where Becky sometimes helps out. Together with friends, she's set up the Food Surplus Cafe – a group of volunteers who collect up the leftover food from local cafes and shops, and use it to cook a meal for the community. Last time they held an event, they fed over 250 people.

The tops of the trees are tinged with morning sunlight; the rush-hour traffic is starting up on the roads.

'This way,' Becky says, and strides across the green. The slope dips down again into a boggy wood.

It's quiet in here. A stream trickles along the valley bottom; birds fidget in the trees. We pick our way past brambles and trails of hanging ivy. Every now and then we hear creatures scuttling through rotting leaves.

Somewhere near the heart of this valley – this dip of land beyond the golf course, between a housing estate and a hospital – a colony of bumblebees has nested in the dried-out carcass of an old tree. The tree stands upright, bleached and skeletal; the nest is hidden inside the trunk. I saw it once when I came here with Jack, but we can't reach it today. At this time of year the path leads only so far, and then it's lost to a bed of nettles. We try to hack a route through, then give up and flop down instead on a fallen log.

I've been describing to Becky the process of honey-making – how flower nectar is collected by bees from miles around and brought back inside the hive where it's regurgitated and passed around to draw the moisture off, before being deposited in the wax cells and sealed as honeycomb.

'Wow,' she says, pulling the thermos and two plastic-wrapped airline-supply biscuits from her bag, 'that's quite a process.'

'Yeah,' I say, taking a biscuit. In fact this whole experience of beekeeping – of learning and asking questions about the hive – has been nothing if not a process, and I wonder if such a process is ever finished; if it ever reaches a conclusion or recognisable outcome. I crinkle the cellophane wrapper between two fingers and suck the sugar from the biscuit, coursing back in my mind over the past year. Suddenly I'm remembering a question I had, when I was first considering getting a hive. What is the link between beekeeping and therapy – and what does honey have to do with healing?

When you type *honey + healing* into a search engine you get a lot of links to health-food websites and some articles about honey's medicinal uses, benefits and side-effects. *More studies are needed to decide if it is safe and effective for various medical conditions* states one website, a little ominously. Another urges me to *consume honey responsibly.* The list of ailments honey is said to alleviate is long. It includes chesty coughs, wounds and burns, insomnia, MRSA, dandruff, lung and liver problems, heart problems, gastrointestinal problems, and memory function in postmenopausal women.

In fact the accounts of honey's use in human medicine date as far back as records go. A Sumerian tablet from around 3000 BC recommends it for skin ulcers. Honey was common in the medicines of ancient Egypt, China, India, Greece and Rome: used as a treatment for both internal and external complaints, it was prescribed to heal eye inflammations, treat mouth and skin sores, and cure chest and stomach troubles.

In *Bee*, Claire Preston writes about a verse in the Qur'an, which describes how bees were instructed by God to *feed on all kinds of fruit . . . From their bellies comes a drink of different colours in which there is healing for people.* It's true that honey comes from *all kinds of fruit*, or flowers, and this means that – just like the bees – it is not easy to categorise. A bee can't be thought of like a cow or a silkworm, since honey is not like leather, milk or silk; the honey doesn't come from her body. She creates it from raw materials external to herself, and *together* – mouth-to-mouth, within the colony. This in-betweenness may well have been part of what made honey so attractive as a component of healing rituals and magic rites: it is so difficult to *place*, neither wholly one thing nor the other; not animal (since it isn't extracted from the bee herself); nor completely plant-based (since it is processed by the colony). It is gathered from countless flowers and flower species, and produced by thousands of individual bees.

I'm searching for milk in a health-food shop when I catch the tail end of a conversation between three men standing over the honey section.

'A spoonful a day you're supposed to have.'

'On its own? I don't like it on its own.'

'It's good for hay fever, my cousin says.'

'Local, is it?'

'Lavender.'

'Christ – have you seen the price of this one? What the fuck's mānuka?'

They amble on down the aisle, and I edge my way around a stack of oatcakes to see the honey selection for myself. There is not just Local, Lavender and Mānuka; there is also Raw, Active, Organic, Unfiltered, Hilltop, Wildflower and Acacia. So many different health-food honeys that I'm not sure how anyone is able to tell what's good for them.

I come across a study by researchers in Switzerland, who tested honeys from around the world for traces of five common neonicotinoid pesticides. Of the 198 honeys sampled, 75 per cent were found to show traces of at least one; 10 per cent contained four or five. The colony's own food supply is being contaminated – which turns the picture of honey as health food on its head.

Snapping the laptop and the research study closed, I glance out of the window at the hive. It's almost a month now since I last opened it up, but as August draws to a close I've begun watching it more closely again. If there's honey in there, what will it be composed of, and what might I become composed of if I eat it?

Today the flight paths are heading south, in the direction of a warehouse roof and beyond that who knows. In reflective mood, I take my boots and head outside; take up my position on the stump.

I'd imagined that in getting a colony I might come to know the bees, understand something about what they are and how they're linked to what's around them. Perhaps, I'd thought, there might be something therapeutic about that. But the actual experience of a thing is always different from your imagining, and with the foraging season nearly over I'm not sure I understand them any better now than I did before. In fact, when I think back over the last few months, I've more often been struck by the sense of confusion and unsettledness waiting underneath the lid than any recognisable order. It's been the same outside the hive, where the bees are all the time flying out, and you can't tell where to, or exactly what they're bringing back – so much so that it often seems impossible to define anything in its entirety.

The wind rustles the leaves in the hedge, and I shiver. There's a chill in the air, that first feeling of summer ending where the air is different but the light isn't yet. I watch the bees drifting out from the hive

and imagine those light-sensors on their heads, searching for it.

I won't ever hear what they're saying to each other,
but sometimes I've had the feeling that what's happening
inside the hive is speaking to things inside me all the
same. I've tried to keep a healthy distance – to become
a careful and rational observer of the bees, capable of
seeing clearly – but at times it's felt like a boundary has
slipped, and the details I am noticing about the hive are
in fact the ones I most need to notice about myself. Or
it is the other way around, and when something shifts in
the hive it also begins shifting in me, so that the bees are
not just bringing things up, they are also affecting what
happens, which is about as far from rational and detached
as you can be.

A finger of bindweed has wrapped itself around one
leg of the hive and now I crouch down, crawl under,
tug it out. Pulling myself up, I look back for a moment
at the house. There's a sense of activity and inhabitation
about it now that wasn't there before. The pots outside
the kitchen window are brimming with nasturtiums; the
back door is propped open most days by a rusted colander.
There's been a change here. It's been happening while I've
been sitting on my tree stump, not doing anything very
much. While the bees have been busily getting on with
life. I struggle to locate what it is, this change, and maybe

anyway it's still in process. The beekeeping season isn't over, and we haven't got to the honey yet.

The next day I arrive home to find a large plastic box on the doorstep. I'm tired and hungry and the box is in the way so I climb around it, assuming it is something to do with Becky. But when Becky gets home later and climbs around it too she says that it isn't. We go out for a closer look. It is not just a plastic box. There are six white tabs at the top and a wooden rack fitted inside the lid. There is no note attached, and we don't know who it's from.

A while later I get a call from Jack. 'Did you find it?' he asks excitedly. 'Can you guess what it is?'

No, I say, we nearly threw it in the rubbish bin, and he sounds a bit surprised and disappointed that we haven't recognised his new invention: a device for honey-harvesting.

Techniques for honey-harvesting depend a lot on hive type. With a modern hive you take the frames out and uncap the wax-crusted honeycomb with a thin-toothed comb before fitting the frames into an extractor – a large, cylindrical spinning machine that spins so fast that the honey flies out and then drips down to collect at a tap at the bottom. The frameless comb from a top-bar hive would fall apart if it was spun, so instead the comb is removed from the hive and sliced clean from the bar before being strained through a sieve. It all sounds quite straightforward, except for one fairly major complication: how to separate the bees from their honey? They don't give it up easily, and will even seek to claim it back where they can. So Jack has been working on a solution.

'You put a piece of harvested comb in the box,' he explains, 'and then you put the lid on. Did you see the white valves at the top? They're one-way exits. Once the bees climb out they can't get back in, so you sit and wait until they've all come out; eventually there's just the honeycomb left.'

'Wow,' I say, suddenly impressed.

'So shall we set a date? For the harvest, I mean. September's just around the corner – it must be about time.'

I had been thinking that learning about beekeeping was a bit like the experience of a foraging bee, gathering scraps of knowledge and insight from all around like the pollen and nectar she collects and carries back into the hive. Now I'm wondering if it's more like *trophallaxis*, that process of communal sucking and spitting that happens inside the hive when the nectar is there but hasn't reached the honey stage yet.

Conversations this year have fed into each other. New insights prompted others, turned old ones around, moulded new shapes out of partially digested forms. There were the conversations with Luke on the phone, and Huber's letters, and the friends who came and challenged what I was seeing. There were also journeys out, to the ONBG meetings and the museum, and to that flat of his in Forest Hill that now has become a place I keep going back to.

It is possible to gauge the range of communal digestion inside the hive by treating a sample of nectar with a radioactive tracer. James and Carol Gould describe how, if the nectar is carried into the hive in the morning, by evening almost the whole colony will be marked; nearly all the workers take part in its conversion into honey.

'Hold *still*!' Pat says, because I've been so busy talking I haven't noticed that the coffee in my cup is about to spill. I can say his name now; I don't have to call him the psychic any more. I'm in London to see him the weekend before the honey harvest. We're sitting on a pavement outside a bakery eating day-old pastries, a late breakfast or an early lunch on our way to see a friend.

He reaches over and takes a swig of the coffee. I make a swipe for his pastry, and miss, and he laughs and hands it over anyway.

I've noticed we're getting bolder and louder around each other. We laugh often and a lot. It is some great relief, I've found, to feel a want and be able to run with it. To have it welcomed, warmed, heated, spurred; to spur and welcome a want in him.

On the pavement beside us lie my notebook and a pen. We've been drawing pictures and writing some things down about What Might Happen Next. We've begun thinking we might want to make some changes, move a bit closer to each other, sometime soon. It is not quite a plan yet; it is just like playing with shapes. And it feels fresh and frameless and deeply good.

I look down at the scribbled notes, which are really the beginnings of a conversation about what home is or could be, and realise there's a story circling somewhere in the back of my mind about honey and healing – I read it in Preston's *Bee* last week.

During the Second Balkan War, the Bulgarian army ran out of medical supplies. The fighting kept happening, and the wounded soldiers kept coming. Inside the medical tents, no one knew what to do; the cupboards were empty, there was nothing left to use. Then someone thought to try honey. They began using it under bandages, as a wound dressing, and it worked. Honey's high osmolarity means that it is capable of drawing moisture off surrounding cells; it can kill bacteria, which is useful for preservation inside the hive and also makes it an effective antiseptic.

But there was something else, too. Another property belonging to honey that made it an especially good choice as a wound dressing: it stays liquid over time. So it didn't solidify under the bandages; it didn't tear the flesh as it healed.

Thinking of it now, that story seems to me like a whisper or a promise of something new. That fresh skin can form, new life emerge – that it is possible to feel protected and held and yet still able to move.

I lean back, resting my elbows against the pavement as I stretch my legs into the street. He reaches forward, finds the gap of skin between my shoe and trouser leg, and wraps a hand around it.

Three birds fly over and a rubbish truck turns.

'Shall we move?' he says, and pulls me up.

It's funny what happens when you ease up a bit; it seems to prompt others to follow suit. In recent weeks people at work have softened, even opened up, and those tough exteriors don't seem so impenetrable any more. Still, I'm beginning to see the truth in what Dulcie said: some environments are toxic to certain life forms – and sometimes a bad fit is just that. I'd like to say that I've found ways of adjusting, even making it good – but I'm coming to realise that the best thing in this instance may be to cut my losses, and – movable thing that I am – move on.

We throw our empty wrappers into a bin and walk up the street to a park and playing fields. There are people, pushchairs, a football. Above us the trees are filled with layer upon layer of dry and shifting leaves, and we look up as the light splits through the spaces between them.

# September

On the day of the harvest there are four of us arranged at different distances from the hive. Or there are three of us, with the side-gate open – Ellie arrives when we've already begun, dropping her rucksack by the fence and pulling on a borrowed suit before slipping in beside us.

Jack and I are standing on either side of the roof; Ellie and Becky are on the grass beside the plastic box, which is filled with our collected equipment. The equipment consists of a hive tool, a bee brush, a large knife, an assortment of spoons and forks, four sieves and a large metal pan.

'Ready?' Jack says.

'Ready.'

A low buzzing rises as we lift the lid. Jack asks Becky for the hive tool but Becky doesn't know what a hive tool is, so Ellie picks the box up and brings it closer – but then she's blocking the entrance and the bees get flustered about it.

'Careful,' I say. 'Not the flight path.'

'Oh – sorry.' We're all a bit over-polite and unsure of ourselves, and I begin to wonder if we should have planned this better.

'Let's start at the back?'

'Here?'

'Got it?'

'Yep.'

'Ready. Go.'

It is clear right away that the inside of the hive has been transformed. Where eggs and larvae lay only a month ago, we now find row upon row of honeycomb. Perhaps *row* isn't right. The comb is more like waves. It undulates, making rounds, not lines. In parts where the honey-making process is still incomplete we see liquid shining in the uncapped cells. Elsewhere there are swathes covered over with a light wax skin. The skin is so thin you can see the honey through it, and it's white and crisp where it touches the cell walls.

'Look here – there's more.' The whole substance of this place has changed. I'm tempted to stop what I'm doing and just marvel at what's happened here, but the bees won't put up with us for long so for now I keep my wonder to myself. We work our way through the bars one by one, and find six filled with honeycomb.

'That's got to be enough for a harvest,' says Jack.

'Shall we?'

'How many?' We take two; and the bees cling to them. Jack shakes each bar once, hard, and a ream of bees falls down and lifts up around us.

A honeybee will make around one-twelfth of a teaspoon of honey in her whole lifetime. If you were to cut a piece of string equivalent to the combined flight distance made by foragers for just one jar of honey, it would reach almost one and a half times around the world. As I carry the comb away from the hive I wonder how many lifetimes this is; how much is it okay to want?

At the back door Becky is waiting with the bee brush and we sweep the bees off like dust, except we keep getting legs caught in the brush hairs. When only a few bees are left we place the comb inside the plastic box and fit the lid on; the one-way valves work just as predicted, and we watch as the bees crawl out one by one.

Back inside the house we arrange ourselves in a kind of circle, but the circle is crooked because of the furniture. Jack is squashed between a chair and a lamp stand; Ellie has one leg under a table. Becky's begun plucking open the wax capping with a fork, and now the rest of us lean in. Inside each cell is a droplet of honey, and inside each droplet is a spot of light from the window. Jack sticks in a finger and licks it.

We take the sieves and hold them over the pan, then begin straining the comb and mashing it with our spoons. The honey gloops. We've all begun licking our fingers now, which makes us jittery and a bit rushy too. The taste makes me think of something high up and intricately spun, clear and sharp and full of the sun.

'I think I'm getting a headache,' Becky says. And I can't shake a feeling in the room of intensifying light. Perhaps it is the smell. The smell is like sweetness and wax and splintered wood.

Soon the honey is everywhere. There is no stopping it. I abandon my spoon and begin using my hands. My hands glisten and stick to things, and everything sticks to them.

Then, quite suddenly, there's a shift. It is no longer just us in the room; something else, dark and shadow-like, is flitting against the walls and around the edges of our vision. 'A bee!' Ellie cries, and I look up. It scrapes the ceiling and dives back down. 'There's one here too!'

We look into the corners, around the walls. They must have come in with the comb, we say to each other. But the one-way valves seemed to work, and anyway the numbers in here seem to be increasing. There's another on the pan lid; one resting on a sieve.

'We've brought brood inside,' Becky says, and it's true. In my top bar hive there was no way of separating the queen from the rest of the colony, so the brood was not separate from the honey stores; and the comb we harvested contains eggs and larvae too.

'They're hatching!' Right here, in our hands and between our sieves and spoons. I peer down at the sticky, broken mess in my hands. There are bodies in there, scratching at surfaces, pulling up.

We stop, stand up as if to do something, though none of us knows quite what to do.

'But they can't be flying,' says Jack.

'Why can't they?'

'I don't think they can, that young. We must have brought them in with us.' And with a frown he begins inspecting the one-way valves on the box lid. 'The guy in the shop promised they'd work,' he says crossly, putting one eye up to a valve and squinting.

We caught the ones we could, and took them outside. And then we finished straining the honey. We filled one large Kilner jar and six small ones – it took us three hours.

'How did it go?' Pat said, when I called up to tell him about the harvest.

I tried to describe what the inside of the hive looked

like and how we harvested the comb, what it felt like to carry it away in my hands. Then I told him about the hatching bees – how it happened all of a sudden, there were bodies in there as well as honey, I was holding birth and death and harm and care all at once, all on top of each other, and then there were also the live bees strimming the walls.

I said it all in a bit of a fluster and the words didn't come out right, I got muddled or upset and anyway I am never very good at describing things over the phone.

'I want to see you,' he said.

'I want to see you too.'

How much of looking, how much of wanting to look, is about its opposite – about wanting to be seen?

Later that night we wiped everything clean – the floor, our hands, the outsides of the jars – but there must have been a residue because a slight tackiness lay over everything for a while after. I could still feel it this evening, nearly a week on. I stepped on a floorboard, and the floorboard stuck.

I wrote to Paul to tell him about the live bees that flew up as we strained the comb, and asked where he thought they might have come from. He wrote back with three possible explanations, laid out in a neat, bullet-pointed list.

They could well have been newly hatched bees; when bees first emerge they do, after all, have wings. Or it's possible that some bees found their way in from outside; some beekeepers say that if you're processing honey within proximity of the hive, the bees will always find a way of reaching it. Or, as Jack had thought, perhaps the one-way valve went both ways.

So it seems that the case of the flying bees in the living room is set to remain a mystery, something we'll never quite make sense of, and maybe I even like it better that way.

We kept one jar of honey for our house. The rest I'll carry out, lids tightly screwed, and give to friends. I took one when I went to London at the weekend, where there was a gathering of all those friends who put money in last Christmas to buy me the colony. No one wanted to open it at first. They passed it around and looked, but it wasn't until the next morning that Laurence made a pile of toast and then the lid came off and everyone tried it.

I gave one to Pat, who keeps it on his desk and has a spoonful with each cup of coffee. There is a particular time each day, he tells me, when if the sun is out it will hit the jar and send a pattern of refracted light against the wall – an amber and orblike glow.

I offered Ellie and Jack a jar each as they were leaving after the harvest, but Jack shook his head.

'Not for me,' he said.

'You don't like honey?'

'I do like honey,' he said, 'but I was wondering. Could I take the wax?'

He wanted to make mead with it. I popped over this week, and he'd boiled the wax down and taken the last of the honey off. There's just enough for one batch, he said. In his bedroom was a row of demijohns filled with fermenting liquids. Beech-leaf wine, beer, blackberry wine, mead. The mead was a dull-dark brownish colour, like dirty pond water. The jars were lined up under his window and along the radiator. When the heating came on there was a soft popping of bubbles as the carbon dioxide escaped from them.

# October

Sometimes, often, I find it easier to make sense of a problem when I can step away from it slightly, view it sideways or through other things. I must be trying to make sense of something now because here I am with the dictionary in my hands again, still trying to tease out the relationship between beekeeper and bee, human and wild creature. Man and his own crazed, tortured desire to keep.

This time I try making it concrete; I look up the *noun* form.

A *keep* is a bar of soft iron placed across the two opposing poles of a magnet to prevent loss of power. A *keep* is a ring worn to hold another on the finger. A *keep* is a piece of wood or plastic inserted through the earlobe to keep a piercing open. A *keep* is the innermost tower inside a castle.

These fortified towers were built deep within the castle walls, and first appeared in Europe during the tenth and eleventh centuries. They served a dual purpose, housing the lord's private residence and also providing a place of retreat should the rest of the castle fall to invaders. Early keeps were made from timber and hung with animal skins to delay or dampen fires; later they were built from stone.

I'm sitting on my bedroom floor, books and laptop strewn pell-mell across the carpet. I'd planned to spend this afternoon clearing out cupboards, sorting through piles of books – but here I am getting into them again.

I wonder about those keep towers, the safe-holds and look-outs hidden inside castle walls. My own need for such places of sanctuary can make me uneasy; I worry that a tendency towards retreat takes me from the world, that I am too prone to withdrawal, too inclined towards places of safekeeping. But perhaps it is possible to see it differently.

I do a bit of reading about the etymology of the castle keep, this time steering away from the dictionary and looking to other sources. Keep towers were also referred to as *donjons*, which was later corrupted to *dungeons* – a term that originally referred not to a place of *imprisonment* but of *refuge*.

Over this past season, it seems to me that the hive itself – that teeming space by the far fence – became a refuge of sorts, granting me the space and legitimacy to begin testing out some different ways of doing things. Down by the hive, away from the city's tough exteriors, I found a place where I could take my armour plating off; could become more exposed, more capable of touching and being touched. Perhaps I became better at caring, too.

I reach up to my bedside shelf and take down the piece of twisted and torn-off comb, which has faded and dried out, the wax melted in places from the sun.

Huber and Langstroth, those pioneering beekeepers, each grappled in their own ways with that sometimes permeable membrane between inside and outside world. Huber was blind, and Langstroth sometimes became mute. The hive seems to have offered refuge to them both; to have been a place where sight and voice and sense-making found a lucidity and a clarity they struggled to find elsewhere.

Last year I was also feeling blocked, caught in a culture and a state of being that seemed to be short on care and to have little patience with sensitivity. The hive, for me, was about escaping that site of difficulty; or the hive was not about escape at all, but about the upwards thrust of my own hard-fought belief that something *else* was possible – a different kind of perception, of relation – within this less-than-perfect range.

\*

In theory, honey keeps. Its high osmolarity forces the water out of yeast and bacteria cells, killing them and preserving the honey itself. Our honey, on the other hand, is unlikely to last long. We've been heaping spoonfuls of it over porridge in the mornings; fragments of wax that escaped our filtering process float to the top, glistening in the milky sludge.

The jar is already less than half-full and I've been half-wishing I'd kept more back, but so much of what I learned as a beekeeper this year came through other people that it seems right to pass the product on. I've saved a jar for Charlotte who took me to the natural history museum, and I took one into work. When people are handed a jar, I've noticed that they often hold it up to the light. It seems almost automatic, and not dependent on source – I gave one to Hannah and she held it up to an electric light bulb.

Two hundred and fifty grams went into a honey cake. I'm going to Wales with Pat at the weekend, the first journey we'll make together, a change from shuttling back and forth between my house and his. I thought a cake might be a good thing to take. A friend of his bought a derelict Victorian schoolhouse a few years ago, and has restored and reopened it as a painting school. He's putting a gig on to celebrate, and has asked friends who are musicians to come and play.

I left the cake in the oven too long and the crust came out as black as coal dust, but the inside is soft and sweet and so dense with honey that perhaps it is all the better for a little bitterness on the outside. Becky and I tried some just to check it was edible, and then Ellie came over and tried some too. It is a big cake, so there is still a lot left. I just hope that no one in Wales minds that it is less than whole, or that there are burnt bits on the crust.

I mentioned to Ellie where we were going, and she told me about an old abbey nearby that we should go and visit. I looked it up and was redirected to a website for the organisation that looks after Welsh Heritage sites. Its name is *Cadw* – the Welsh word for *keep*.

I won't open the hive again now until the spring. You have to leave it closed through winter so as to help the colony preserve its heat; even the most interventionist beekeepers know this. So this week I packed my beekeeping equipment away for the final time this year, and as the temperature outside continues to cool I've noticed we are also going out to the garden less.

It's Saturday morning and the back door is open; I'm balanced on the doorsill. I tip forward, and my toes tap the brickwork paving outside. Tip back, and my heels hit the stone tiles of the kitchen. We leave for Wales in twenty minutes and Pat's upstairs deciding whether to take soft shoes or walking boots, and which is the best music for the car. I'm wearing wellies, the boot sawn off at the ankle to make them lighter and easier for clambering. My bag is packed and waiting by the door, the honey cake wrapped in foil and balanced a little precariously on top. We ate a piece each when he arrived last night, just to check it was still fresh. (It was.)

I look over at the hive and spot a bee lifting from the entrance. It's been raining for days, and this is the first one I've seen all week. I step out of the doorway, walk down to the far fence. With wintry weather forecast, this might be my last glimpse of them for a while.

At the base of the hive a small heap of shrivelled-up bodies lie scattered across the concrete block. These are male drones, expelled from the hive and left to die of starvation rather than consume any more precious resources. The colony is ridding itself of surplus, reducing down to a core body of workers and the queen until spring arrives, when she will begin laying and the colony expanding again.

The ONBG has reduced down for winter, too. I went along to a meeting earlier this week, and we were only a slim handful. But Paul says that as a group, we're thriving. He has a long list of people ready to join next spring, and is even planning a programme of trips in the summer. He's met a lady who runs a plant nursery specialising in bee-friendly flowers – she grows them herself, from seed and without pesticides; she's going to show the group her test plots, and how to monitor which plants are most popular with the bees. There are already twenty-six people signed up to go.

I knew better than to give honey to Luke – he's swimming in it by this time of year; but I did go and see him when I was in London last. 'So you're a real beekeeper now,' he said. 'You've made it through a whole season with a hive.' He's busy at the moment, working on a project aiming to create 'pollinator corridors' in areas suffering the effects of fragmentation. Using existing railway lines, he'll be working with station staff and community groups to build a network of flowering habitats across London and beyond, doing the job of reconnection in places where pollinators are struggling. He's just bought a retro caravan, too. A friend of his is going to convert it into a bee-mobile, a sensory-education centre for kids. There's work coming in already and they need more hands. He asked if I'd like to get involved, and I said that I would.

Three more bees emerge from the entrance, pace the landing board and fly out. That too-bright tangerine colour has softened over summer, faded and streaked by sun and rain, and it doesn't look so out of place in the garden now. Sometimes when I've opened it up I've found spiders nesting in the lid.

'Are you ready?' Pat calls, and I look back to where his head has just appeared from the doorway. 'I'm ready,' he says, 'whenever you are.' He disappears back inside. I notice he's wearing soft shoes; and he's taking walking boots inside a bag.

How to shake the feeling I have when I look at him sometimes that he is not separate from the hive? That through this experience of beekeeping, of learning about and listening to the colony, I might have called something up – might have begun to articulate and name a capacity I was missing, a connection I needed? A particular kind of sensitivity, a quality of attention which is so thick in him it is almost like a substance itself, and which he encourages and bolsters in me. What to *do* with a feeling like that – which is not rational, and doesn't fit with the usual categories – except to notice it silently and with a sideways grin as it becomes part of my day-to-day?

It's time to go. I make a movement towards leaving and then pause a moment and look up. I'm still trying to guess where the bees' flight paths lead to, or even which direction they're headed. And of course I can't. I lose sight of each one somewhere near the fence; they're too small for my human eyes, or my eyes are too far from them. But still, here they are, arriving back. Legs thickened with pollen, bellies tightened. I watch another worker leave. From hive to tree to garden fence, beyond. I turn, hearing the car engine start on the street outside, and walk back to the house.

# Acknowledgements

This book is born of the many encounters, conversations and exchanges that took place over the course of that year of beekeeping, and I'm grateful to all those people who appear on these pages and to the many others who have flitted about the edges of them. Thanks especially to Luke Dixon, and to Becky Ayre, who lived this story at close quarters, and coped brilliantly.

I am indebted to Paul Honigmann of the ONBG and Oxfordshire Beekeepers Association, and to Jack Pritchard, each of whom tirelessly answered my queries, untangled my confusions and scrutinised this manuscript for accuracy. Such patience and generosity of spirit is something truly special; and they never once (to my knowledge) laughed at my silly questions. Any inaccuracies that remain are my own.

# Acknowledgements

Thanks to the editors at the *Junket* who published an excerpt of this book at an early stage, and to Skaftfell Centre for Visual Art in Iceland and the Alde Valley Spring Festival in Suffolk, England, who awarded residencies to support the project even when I was still struggling to define what it was. Luckily, there were people around who kept me finding out, and I'm deeply grateful in this respect to Rob Macfarlane for steering me back to the place where it was pulsing; and to Ellie Stedall, Naomi Booth, Dulcie, Laura Joyce, Tom Houlton, Tom Bunstead, Kieran Devaney, Camilla Bostock and Katrina Zaat for helping me to stay with it. Heartfelt thanks also to my agent Jessica Woollard at David Higham Associates for picking it up and carrying it into the world with such care and subtlety of insight, so much vitality.

A full-bodied salute to Rowan Cope at Scribner for her steady and sensitive eye, her call for clarity over evasion. Thanks also to the rest of the teams at Scribner and David Higham for taking this on, having fun and running with it.

For offers of room and shelter during the writing, thanks to Julia Blackburn, Tim and Kate Ottevanger, Michael and Marcia Blakenham, Tom Merilion and Ashley Mellet.

And, finally and above all, thanks to my mum and dad, and to Pat, who opened the windows and brought in the light.

# Bibliography

My learning about bees and beekeeping came from a variety of sources, and a special debt is owed to the following authors and books:

Crane, E., *The World History of Beekeeping and Honey Hunting* (New York, NY: Routledge, 1999).

Ellis, H., *Sweetness & Light: The Mysterious History of the Honeybee* (London: Sceptre, 2004).

Gould, J. L. and Gould, C. G., *The Honey Bee* (New York, NY: Scientific American Library, 1988).

Preston, C., *Bee* (London: Reaktion, 2006).

Wilson, B., *The Hive: The Story of the Honeybee and Us* (London: John Murray, 2004).

François Huber's letters were originally published in 1792 in a single volume; a second edition was published in 1814 in two volumes, edited in part by his son Pierre Huber. *New Observations Upon Bees*, published in 1926 by the *American Bee Journal*, is a careful and very readable translation by Charles Dadant.

Two particular texts provided some further background on Huber: *The Life and Writings of Francis Huber*, written by his friend Auguste de Candolle and published by the *Edinburgh Philosophical Journal* in 1833, and Sophia Bledsoe Herrick's *Sketch of the Life of Francis Huber*, published in *Popular Science Monthly* in 1875.

In addition to those mentioned above, the following sources were particularly useful references on bees and beekeeping, or prompts for further thinking:

Abdel Haleem, M. A. S. (trans.), *The Qur'an* (Oxford: Oxford University Press, 2010).

Aristotle, *Generation of Animals*, trans. Peck, A. L. (Cambridge, MA: Harvard University Press, 1942).

—— *Metaphysics, Volume I: Books 1–9*, trans. Tredennick, H. (Cambridge, MA: Harvard University Press, 1933).

Berger, J., *And Our Faces, My Heart, Brief as Photos* (London: Bloomsbury, 2005).

Butler, C., *The Feminine Monarchie* (London, 1609).

Clarence, C., 'A Closer Look: Sound Generation and Hearing', *Bee Culture: The Magazine of American Beekeeping*,

22 February 2016, accessed via www. beeculture.com

Columella, L. J. M., *On Agriculture, Volume II: Books 5–9*, trans. Forster, E. S. and Heffner, E. H. (Cambridge, MA: Harvard University Press, 1954).

Couvillon, M., Ratnieks, F., Schürch, R., 'Dancing Bees Communicate a Foraging Preference for Rural Lands in High-Level Agri-Environment Schemes', *Current Biology*, 24, 11: 1212–15, 2014.

Dade, H. A., *Anatomy and Dissection of the Honeybee* (Cardiff: International Bee Research Association, 1994).

Dalby, A. (trans.), *Geoponika: Farm Work* (Totnes: Prospect Books, 2011).

Dixon, L., *Bees and Honey: Myth, Folklore and Traditions* (Mytholmroyd: Northern Bee Books, 2015).

Dunbar, J., 'Some Observations on the Instinct and Operations of Bees, with a Description and Figure of a Glazed Bee-Hive', *Edinburgh Philosophical Journal*, 3: 143–8, 1820.

Eliade, M., *Myths, Dreams, and Mysteries*, trans. Mairet, P. (Glasgow: Collins, 1968).

—— *The Sacred and the Profane: The Nature of Religion*, trans. Trask, W. R. (New York, NY: Harcourt Brace, 1959).

Giggs, R., *The Rise of the Edge: New Thresholds of the Ecological Uncanny & Inside Albatross: Fictions for Strange Weather* (unpublished doctoral thesis, University of Western Australia, 2010).

Hartlib, S., *The Reformed Commonwealth of Bees* (London, 1655).

Hine, D., *The Crossing of Two Lines* (Stockholm: Elemental Editions, 2013).

Horn, T., *Beeconomy: What Women and Bees Can Teach Us about Local Trade and the Global Market* (Lexington, KY: University Press of Kentucky, 2012).

Langstroth, L. L., *Langstroth's Hive and the Honey-Bee: The Classic Beekeeper's Manual* (Mineola, NY: Dover, 2004).

Pliny the Elder, *Natural History, Volume VI: Books 20–23*, trans. Jones, W. H. (Cambridge, MA: Harvard University Press, 1951).

Seeley, T., *Honeybee Democracy* (Princeton, NJ: Princeton University Press, 2010).

Snodgrass, R. E., *The Anatomy of the Honey Bee* (Ithaca, NY: Cornell University Press, 1985).

Tautz, J., *The Buzz about Bees: Biology of a Superorganism*, trans. Sandeman, D. C. (Berlin, Heidelberg: Springer-Verlag, 2008).

Virgil, *Eclogues. Georgics. Aeneid: Books 1–6*, trans. Rushton Fairclough, H., revised by Goold, G. P. (Cambridge, MA: Harvard University Press, 1916).

Wilson-Rich, N. W., *The Bee: A Natural History* (Lewes: Ivy Press, 2014).

Winston, M., *The Biology of the Honeybee* (Cambridge, MA: Harvard University Press, 1991).

# Further Information

## Becoming a beekeeper

You don't need a lot of space to keep a hive; Viktor managed to keep one on the balcony of his apartment in Kiev. Wherever you live, if you're thinking of getting a hive, it's important to practise in a way that will support both your own colony and other pollinators in your area.

The British Beekeepers Association's *BBKA Guide to Beekeeping* and Alan Campion's *Bees at the Bottom of the Garden* offer clear and manageable introductions; and if you're interested in top-bar hives (I recommend them!), *Top-Bar Beekeeping: Organic Practices for Honeybee Health* by Les Crowder and Heather Harrell, and *The Thinking Beekeeper: A Guide to Natural Beekeeping in Top Bar Hives* by Christy Hemenway are particularly good. Readers in Australia or New Zealand may also like to try *The Australian Beekeeping Manual* by Robert Owen.

The local climate, weather systems, topography and patterns of land use will determine the kinds of issues your colony is faced with, so as a beekeeper it's important to know your area. Look out for a good local beginner's course, and join a beekeeping group; get a mentor who will be able to share their knowledge with you.

Simon the Beekeeper and Thorne are good stockists of hives and beekeeping equipment in the UK. Elvin is still making hives, and has now branched out from his shed – his company Major Beehives comes recommended.

## Building pollinator-friendly habitats

Pollinators of all species urgently need more flowering and nesting habitats to be made available, and there are many ways that you can help make this happen. Grow more flowers, shrubs and trees; cut grass less often, and leave nests well alone; avoid using pesticides in the garden. Try talking to your local school, council or place of work and encouraging them to plant flowerbeds with species that will attract pollinators; get involved in campaigns to reduce pesticide use, and demand transparency as a consumer – ask your garden centre whether they can guarantee that their plants are free from pesticides, or find a nursery, such as the excellent Rosybee Plants, who are growing their own plants pesticide-free. To find out more, have a look at Buglife, Friends of the Earth, the Bumblebee Conservation Trust, or Luke's charity the Bee Friendly Trust – all are doing great work to promote pollinator-friendly planting.